Hazardous Forecasts and Crisis Scenario Generator

Series Editor
Jacques Janssen

Hazardous Forecasts and Crisis Scenario Generator

Arnaud Clément-Grandcourt
Hervé Fraysse

ELSEVIER

First published 2015 in Great Britain and the United States by ISTE Press Ltd and Elsevier Ltd

ISTE Press Ltd
27-37 St George's Road
London SW19 4EU
UK

www.iste.co.uk

Elsevier Ltd
The Boulevard, Langford Lane
Kidlington, Oxford, OX5 1GB
UK

www.elsevier.com

Notices
Knowledge and best practice in this field are constantly changing. As new research and experience broaden our understanding, changes in research methods, professional practices, or medical treatment may become necessary.

Practitioners and researchers must always rely on their own experience and knowledge in evaluating and using any information, methods, compounds, or experiments described herein. In using such information or methods they should be mindful of their own safety and the safety of others, including parties for whom they have a professional responsibility.

To the fullest extent of the law, neither the Publisher nor the authors, contributors, or editors, assume any liability for any injury and/or damage to persons or property as a matter of products liability, negligence or otherwise, or from any use or operation of any methods, products, instructions, or ideas contained in the material herein.

For information on all our publications visit our website at http://store.elsevier.com/

British Library Cataloguing-in-Publication Data
A CIP record for this book is available from the British Library
Library of Congress Cataloging in Publication Data
A catalog record for this book is available from the Library of Congress
ISBN 978-1-78548-028-7

Printed and bound in the UK and US

Contents

Introduction

In order to introduce the aim of this book, it is essential to highlight risk trends.

The second half of the 20th Century was characterized by a trend of a decline in major risks. Indeed, the end of the Cold War implied an unforeseen geopolitical risk setback. Moreover, cyclical economic moderation changes after big oil price changes, as well as sociopolitical crisis risk moderating, notwithstanding neoliberalism being observed. It is now obvious that this declining trend is over and that these trends have reversed.

Geopolitical risks are rising trends in Europe, Asia and the Middle East; the great recession showed how risks are rising in a financial and economic world that is increasingly complex and connected. Economic cycle theories, such as Kondratieff's long wave theory, faced new problems (China and Russia entering in the capitalist system, deregulation and central banking's new philosophy). Sociopolitical risks are increasing whilst lagging economic indicators are recovering. So, it is no longer valid to calculate measures on risk which disregard upward trends. Disregarding these trends in decision methodology, asset management and portfolio building implies big disappointments in the next crisis. A portfolio optimization based on utility function simply using variance (as the only measure of risk in the modern portfolio theory (MPT)) no longer seems logical nor prudent.

The aim of this book is to show that with 21st Century risk trends, an economic scenario generator including crisis scenarios is essential for avoiding risk undervaluation. Economic scenario generators (ESGs) have been used for years by insurance company asset liability management; now, there are compelling reasons to use economic scenario generators (ESG) with crisis scenarios for many kinds of asset management. The aim of this book is to convince readers that crisis scenario generators could be necessary.

In order to be able to cope with the new challenges of the 21st Century, it is necessary to learn lessons from the past. That is why in order to avoid depression scenarios and the multiple-crisis scenarios of the 1930s, budgetary reflation policies as well as overabundant money creation have been necessary. The Great Recession, indeed, was not a great depression because central banks were surprisingly quick and smart in their innovative monetary policies and truly bold in their actions. Indeed, to multiply a monetary base by two or three is audacious because the success of these actions was absolutely unsure: in retrospect, it seems clear that it was the right decision. Governments were also bold in their budgetary policies that followed in spite of the steady debt increase necessary to finance these strong reflationary actions.

Big base money creation such as this implies an unstable context which has two important implications for asset managers and investors who want to take into account these new settings.

Firstly, they have to define probabilities for a wide range of scenarios (geopolitical experts could be very helpful for this), including best and worst case scenarios. Failure probability in order to avoid another depression in case of a multiple-crisis process over a short period of time is important. Many scenarios can be built on this theme. Moreover, the comparison between the 1930 crisis and the current context suggests that among all the monetary problems discovered during the 1930s, the competition between the US dollar, Chinese yuan and the Euro could be more dangerous in the future than the replacement of the pound sterling with the dollar as the reserve currency. Nevertheless, the worst is never sure and some best-case scenarios can be built on positive action by the G20 (the group of the twenty most powerful heads of state or heads of government that could slowly become a ruling world group) to avoid protectionism, trade war, competitive devaluations and currency instability linked to reserve currency problem. The G20 seems to be the only hope in avoiding this major instability risk that could amplify trade war and protectionism risks.

Secondly, to take into account the next crisis risk and a multiple-crisis process risks, measurement risk processes have to be based on an (or even several) economic scenario generator which gives precise risk measurements. The huge creation of base money induces instability, which increases the probability of a future crisis. The bigger the debt-to-gross national product (GNP) ratio is, the bigger the danger of an important financial and economic crisis. Furthermore, when there are really low interest rates (near zero), it would be very difficult to cure a potential crisis or reduce its size since interest rates could not be decreased even further.

Therefore, Chapter 1 will deal with risk-oriented philosophy, forecast risk-oriented philosophy and processes. Chapter 2 is devoted to scenario building processes, especially those to be used in generators with an emphasis on major scenarios and extreme scenarios. Chapter 3 discusses asset management processes using generator methodologies in order to avoid risk understatement and for optimization; how can assets be managed when crisis probability increases? To settle a crisis process adapted to the current economic period, it seems appropriate to elaborate an investment approach centered around the use of an Economic Scenario Generator (ESG).

1

Risk-oriented Philosophy, Forecast-based Philosophy and Process

1.1. A risk-oriented philosophy and a forecast-based philosophy

1.1.1. *Why a risk-oriented philosophy?*

Philosophy is essential for staying rational in a period when there are so many risks, so many risk measures and so many processes to calculate them such that a logical endeavor has to be made systematically in order to consider all the possibilities. A philosophy for an unsteady and unreliable world is needed to stay rational in the face of geopolitical challenges inducing sociopolitical risks, climate challenges and financial uncertainty. Several elements can be mentioned to illustrate our proposition.

It is obvious to note that, nowadays, economic stability is less secure than it was in the second half of the 20th Century: indeed, in 1989, geopolitical risk diminished in an unforeseeable manner when Reagan, President of the United States of America, with global power, began to deregulate by proposing a "North American accord", the so-called NAFTA, creating a common market between the United States, Canada and Mexico. This decision taken in a globalized world, in which transmission capabilities allowed an online financial integration, had a significant impact on risk and economic stability. In the 21st Century, this new world appears to be unsteady with an unproven reliability; weaknesses appeared with crises that became more frequent and more systemic than previously. Therefore, to stay rational, a philosophy is needed.

A global financial integration, including many developing and emerging economies with recent financial experience with unproven learning capability, implies a global risk supervision that is not available with the current International Monetary Fund (IMF) or other international organizations. The IMF missions and means have to be enlarged, especially with learning lessons for less-skilled countries; its credibility will be increased when important emerging countries are more present and active in this international organization. Up to now, financial governance has been insufficient compared to world financial integration. A last resort of international lending is needed. Risk-oriented philosophy would evolve if these reforms were well made.

Another source of instability is concerned with the status of the US dollar. Indeed, the reserve currency status of this money could be challenged in the years to come because it is criticized for its weaknesses resulting from big money creation and the US trade balance deficit; however, this deficit adds to the dollar's international monetary base which has to be increased regularly to avoid recessions. Is this monetary base well adapted for world growth? This is an example of a problem without any easy solution. A currency war as seen in the 1930s would lead to a very dangerous risk of financial and banking crises. A risk-oriented philosophy is essential for staying rational in a currency war.

1.1.2. *Management by crisis is the philosophy of global capitalism*

A crisis entails necessary government action and some urgent reforms; after the end of the crisis, reforms are less urgent. This applies to every capitalist country and possibly to less democratic countries that use capitalism as a recipe for growth.

The world integration implies reforms. An initial step was the G20's first meeting in 2009 in the middle of the Great Recession, but this was not enough. To make globalization more reliable with a financial integration which is essential to reduce crisis probability, the G20 has to set up the necessary reforms. For that, the G20's legitimacy power, particularly on the regulatory side, has to grow significantly. Taking into account all the financial crisis propagation scenarios and their mean reversing possibilities with good and bad consequences requires a philosophy of capitalism. To stay more rational than competitors, it is essential to study many kinds of possible crisis in order to not be surprised; the best asset management returns can be obtained by this philosophy; contrarian asset managers (assessing higher probabilities than the market to turnaround and the mean reversing process) are able to sell very early when crisis probability increases and to buy early when the mean reversing process probability increases. Usually, these contrarians are very experienced asset managers, often located in Boston or Edinburgh.

1.1.3. *A forecast-based philosophy and risk evaluation processes*

In a crisis-prone period in which there are so many reciprocal action propagations for solving problems, philosophy is essential for staying rational. Many forecasts, in this kind of period, are based on imperfect measures. Many forecasts are needed; however, there are so many hazardous methods for forecasting figures, trends, reversals and so many imperfect measures that some logical rules are necessary. Crisis processes caused by increasing fragilities are so difficult to forecast that an endeavor to improve methods is necessary to make a forecast process suitable for forecasting a crisis period; a methodological thought based on an economic philosophy is necessary. The French *Conseil d'Analyse Economique* (French council of economic analysis) in a financial crisis report [BOY 04] emphasized a new concept for monitoring a crisis: the financial accelerator. As firms are further leveraged, financed mostly by credit, an increased profitability has a strong effect on firm value and investment capability; this effect (and its opposite) increases the size of the cycle; there is also the pro-cyclical effect of the credit multiplier. These instability factors and their correlations have to be monitored, but imperfect and asymmetric information is spread by oligopolistic channels; therefore, risk is taken by economic agents, especially firms in a pro-cyclical manner that increase crisis risk. Herd effects, bubbles induced by excessive liquidity with excessive risk taking, are increasing as financial activities are more pervasive. Stronger competition in banking activities implies more risk taking and leverage. Globally, banks add up all these risks.

In a complementary report to the financial crisis report previously mentioned, Mario Dehove [DEH 03] quotes Stone and Weeks' paper [STO 01] covering 1992–1999; using seven crisis leading indicators and the probit method, 16% of observations were found to be wrong crisis forecasts and two crises out of five were predicted. Crisis forecast is very difficult even with a good statistical methodology.

1.1.4. *One-year scenarios and leading indicators*

Leading indicators and indications are to be used in a logical process to build one-year scenarios. When the future of a boom becomes uncertain, the cyclical reversal comes with an increasing risk aversion which induces operators to be more short-term minded. The horizon changes as mean reversal could come from pro-cyclical activities and regulations. Moreover, *to build a one-year scenario it is essential to stay rational* in a crisis-prone period. Indeed, as Pascal Blanqué, Chief Information Officer (CIO) of a major asset management company, wrote in his book *Essays in Positive Investment Management* [BLA 14], it is necessary to bear in mind that several very useful concepts for modeling in healthy periods are to be called into question when a change of paradigm occurs:

> Some of the holy cows were already being questioned before the crisis, i.e. market efficiency hypotheses, CAPM, the Gaussian law... the crisis has killed some of them – but not definitely.

Professor Roubini [ROU 06] forecast the Great Recession at the 2006 Jackson Hole Central Bankers conference as the US real-estate market became bearish after an excessive bull market ended by the first half of 2006. Some leading indications came in June, July and August 2007 from real-estate credit-linked problems in some funds. These early crisis signals were confirmed by the Californian real-estate credit shadow banking system in the summer, and further in the fall of 2007. A run on Northern Rock in Great Britain was avoided by government intervention; Bear Stearn's failure was solved by a merger helped by the Treasury authorities in the winter of 2007. We can, therefore, say that there were crisis leading indications for more than a year and a half. Moreover, these leading indications were earlier than published leading indicators. Some official real-estate credit amounts published in 2007 were simply wrong. Much bigger and much better figures were published by the IMF during the winter of 2008.

A very useful tool for anticipating a recession is the composite leading indicator of the Organization for Economic Co-Operation and Economic Development (OECD). Indeed, these indicators do not miss recessions, particularly in the USA as it is shown in some ENSAE working papers listed in the Bibliography. Nevertheless, they indicate a recession risk that does not materialize as often as they indicate true recessions. It is an improvement compared to the time when Samuelson said that leading indicators give many more false recession signals than they do right ones. Working papers of mathematical statistics (see Bibliography) specifically studied leading indicators' reliability and showed that OECD composite leading indicators' record qualities. With OECD business tendency and consumer opinion, reliable help is also given to forecasters in many countries and in large areas. McCulloch [MCC 95] showed in 1975 that when a recession is not short, turnaround probability and recession duration are independent variables. Niemira [NIE 91] showed in 1991 that turnaround probability is a variable independent of the duration of trend in force. Many working papers were published on these indicators showing difficulties in making a precise crisis diagnosis.

The Conference Board leading indicator gives a heavy weight to average weekly hours of manufacturing, manufacturers' orders and supplier deliveries; this indicator and the Economic Cycle Research Institute (ECRI) indicator seems to be somewhat earlier than the OECD leading indicator, but they missed some recessions when the OECD composite leading indicator (OECD CLI) did not miss recessions and gave less false signals as studied by Debarre et al. [DÉB 12].

	Optimal months' number of predictions	Non-predicted downturns (%)	Non-effective predicted downturns (%)	Effective prediction (%)
CLI	5	0	17	83
S&P500	6	23	27	50
ECRI	6	21	24	55
Conference Board	7	16	17	67

(Source: [DÉB 12])

Table 1.1. *Effective prediction percentage table*

The National Bureau of Economic Research works on 150 series and uses the 12 best leading series and four coincident indicators for confirmation. Eurostat business climate indicators by industry for European countries are also helpful for forecasters. On the whole, leading indicators are not much in advance of equity markets which are relatively reliable leading indicators of the economy. Most of the indicators have jumps and mean reversal moves so that volatility-based significance tests on hypotheses can help. There are a dozen statistical techniques for forecasters. Turnaround indicators appeared in the 1990s. They are the most helpful.

Bouthevillain and Mathis wrote in [BOU 95a] and [BOU 95b] that the experience of the French National Institute of Statistics and Economic Studies (*Institut national de la statistique et des études économiques* (INSEE)) for the latest downturns was unsatisfactory, even if the downturn indicator, "*l'indicateur synthétique de retournement*", was the state of the art. Short-term forecasters have coincident indicators, but this art is truly difficult:

– the liquidity level variation effect on economy was 3–6 months lagging;

– the relative liquidity level variation effect on two economies implies in exchange rate move lagging by 3–6 months;

– the liquidity level variation effect on precious metals was 3–6 months lagging.

World instability is no help to forecasters; lag could change.

When forecasters are late compared to the market, efficiency of the market is more probable than in the case when forecasters are able to be in advance of the market, which seems to be not as well informed as forecasters. The INSEE did a study on the 2008 crisis experience using three sources [RED 09]: French Observatory of Economics (*Observatoire français des conjonctures économiques* (OFCE)), NATIXIS research and INSEE. This published study concludes that:

– *the recession warning came somewhat later than the reality;*

– *the crisis seriousness measure came much later and recession size was underestimated;*

– *nothing in the December 2008 analysis indicated that a deep recession was on the way.*

The difficulty to forecast some scenarios is well illustrated by the Lehman bankruptcy. Indeed, the Treasury Secretary decided not to put the Lehman investment bank into Chapter 11 for rescue and possible recovery; this decision was not possible to forecast, except by people aware of the personal conflict between Fuld, Chairman of Lehman, and Treasury Secretary Paulson, who had managed Goldman in the past. New York fund managers were well aware of this problem. During the inflationary process of the 1970s, only two economists, Henry Kaufman and Albert Wojnilower, understood the size of the phenomenon, and both were former economists of the Federal Reserve System. In the disinflationary process, very few economists were able to forecast this process as well as the US economist consensus. Generally, macroeconomic liquidity leading indicators in the USA are thought to be the best by New York equity market operators:

– M2 and interest rate spread (10 year Treasury bonds less federal funds);

– total non-financial corporate financing as percentage of GNP;

– change in business inventories % GNP;

– change in consumer credit % GNP;

– change in residential mortgage debt % GNP;

– total government financing gap % GNP;

– balance of current account balance payments.

It happens that the US equity market was the best leading indicator in the second quarter of 2009 because the contrarians from Boston and Edinburgh were right to bet on the US convalescence and a great bull market. The market can also be seen as the consensus view of the most active investors; this consensus can sometimes be better than the economist consensus.

In countries with available statistics, statistical series concerning the labor market are the leading indicators of employment; there are production order books that are leading production by many months because the production process is a time-consuming process, especially for equipment and some intended purchases by enterprises and households. Trade balances are not too difficult to forecast; but currency exchange moves are so difficult to forecast that many people compare them to random walk influenced by unknown capital moves that can be irrational.

To conclude on leading indications, a warning can be emphasized: when forecasts and short-term views' dispersion increase, it can be seen as a warning. Many professional asset managers and most renowned strategists wait to publish warning views and recession-size views because they fear that a mistaken warning could be too negative for their career; the youngest are bolder and willing to gain recognition. Renowned strategists are bolder and quicker to speak about the coming crisis *viva voce*.

1.1.5. *Forecasting ability is limited by exogenous shock risks*

Exogenous shocks (geopolitical, sociopolitical contagious effects, climate-linked problems, etc.) are not foreseeable; they are only *more or less* foreseeable. Experts can even give a probability of occurrence on a defined period of time. However, in an unsteady and unreliable world, information and sociopolitical knowledge do not give much forecasting ability; moreover, exogenous shocks induce fear that induces crises and illiquidity in the markets; excessive behaviors could become totally irrational with herd effects. As a result, a rational expectation hypothesis and efficient market hypothesis are not to be considered in these periods. Excessive reactions to bad news and weak reactions to good news in bear markets make an abnormally noisy market; usually, the Kalman filter can erase, on average, one-third of noises. In the scenario of Middle East war with the oil price shock or in the scenario of a major systemic financial crisis, there would be such a fear and such a stampede to sell that illiquidity would be widespread and confidence would disappear. As it is well known that capitalism is based on confidence, it is clear that forecasting ability will disappear resulting from the fact that:

– social and political conflicts, erupting war, rebounding war in conflict-prone areas, etc., are quasi-impossible to forecast. Many incorrect forecasts are the result of impossible to forecast social events, regulation change or political events (in a working paper entitled *Perspectives politiques et prévisions économiques* for the third cycle of *Institut d'études politiques*, A. Clément-Grandcourt examined the political climate and newspapers' clear-sightedness in limiting forecast ability);

– time is very important in social and political life; everything is obvious some years after the event but not possible to forecast, sometimes incredible; imperfect knowledge, shallow information by media, psychological vagaries and very hazardous learning are the sources of mistaken forecasts.

The Nobel Prize winner, Maurice Allais, in his lectures for engineers and statisticians has studied factors linked to exogenous shocks and reactions. More precisely, he has scrutinized several kinds of reaction: competitors' reactions imperfectly understanding the shocks, effects of wrong forecasts and of wrong reactions for enterprises, households and central banks; effects of inadequate and late information for competitors. A corrective process is generally very difficult to forecast but scenarios, based on experience, can be built:

– time lag to understand corrective actions is not possible to forecast, even if learning processes can be somewhat effective; many scenarios are needed;

– corrections of mistaken political actions cannot be forecast; however, it is possible to give probabilities for scenarios of corrective actions.

The best forecasting ability is linked to demographic forecasts even if they cannot erase uncertainty on fertility rates and migrations flows. Economic forecasts of social cover (social provisions about health, pension fund deficits and the cost of education) can be specifically medium term and not so incorrect in the long term. It is certain that these costs will grow in developed countries so they will increase the crisis severity and the difficulty of exiting the crisis. So, these demographic forecasts are essential. Social indicators are linked to demography; some of them could give some ideas about the future but it is rather indefinite: inequality indicators, precariousness indicators, purchasing power change and quality of life indicators. These indicators, estimated for different groups of people, could help to assign probabilities to social and political crises that have important consequences for the economy; delinquency and insecurity indicator's increases could also warn of a potential downturn. Social crises cannot be foreseen even if there are some leading indicators of social unrest and dissatisfaction.

Modern portfolio theory (MPT) optimization is the best in efficient markets with a rational hypothesis; in periods that are prone to exogenous shocks and crisis, it is very different; the MPT utility function using variance as a measure of risk is not good in a crisis; by trial of other utility functions, backtests on crisis process and corrective action, it is possible to find much more suitable utility functions that would have been truly good during the 2008–2010 period. Therefore, a backtesting process is an empirical method to obtain good allocation for crisis. More precise information on these topics is available in Chapter 2.

1.1.6. *The necessity of scenario building*

In a globalized world, with a very limited forecasting ability, scenario building is required.

Short-term forecasting reliability is satisfactory on usual economic and financial cyclical patterns. Forecasting has limited reliability in the case of unusual patterns and crises; many crises are impossible to forecast even on a one-year horizon. Scenario probabilities can be estimated for the mean reversing process, disequilibrium corrections and crises that are not possible to forecast. Forecast philosophy is different from scenario building philosophy. Forecast is the most probable scenario without any estimated probability; one or two alternative scenarios can be added. Scenario building is an endeavor to cover all possibilities which have an estimated probability higher than 0.1% in order to be able to estimate value at risk (VaR) for 1 and 5% as well as conditional value at risk (CVaR) and tail value-at-risk (TVaR) (see Box 1.1 for more details on these indicators).

In financial mathematics and financial risk management, value at risk (VaR) is a widely used risk measure of the risk of loss on a specific portfolio of financial assets.

Given a confidence level α in [0-1], the VaR of the portfolio at the confidence level α is given by the smallest number "l" such that the probability that the loss "L" exceeds l is at most $(1-\alpha)$. Mathematically, if "L" is the loss of a portfolio, then $VaR_\alpha(L)$ is the level α-quantile, i.e. [MAR 52]:

$$VaR_\alpha(L) = \inf\{l \in \mathbb{R}, \mathbb{P}(L > l) \leq 1 - \alpha\}$$

It is well known that the VaR is a risk measure that is not coherent because of the non-respect of the hypothesis of subadditivity.

Consequently, another risk measure which is coherent is often retained, the tail value-at-risk (TVaR), also known as tail conditional expectation (TCE). This indicator is more general than the VaR and quantifies the expected value of the loss given that an event outside a given probability level has occurred. The TVaR at α level is defined as follow [MAR 52]:

$$TVaR(X; \alpha) = \frac{1}{1 - \alpha} * \int_\alpha^1 VaR(X; t)dt$$

TVaR is then the mean of VaR above the α-level.

Box 1.1. *More details on value-at-risk (VaR) and TVaR*

Medium-term forecasting is limited to some classic scenarios; it is not possible to forecast sociopolitical and geopolitical crises on a five-year horizon, but scenarios can be built; disequilibrium corrections that cannot be forecast can involve many scenarios. Learning from the past is important in scenario building, but to foresee new kinds of scenarios caused by the rising risks of the 21st Century is more important. Many crises are possible to forecast in a global era; scenario building to explore possibilities is the way to rationalized decisions.

1.1.7. *The importance of crisis propagation scenarios*

Crisis propagation scenarios are essential for risk calculations and risk aggregation philosophy.

At the beginning of the 20th Century, a free trade era under the *Pax Britannica* implied a major crisis contagion from Asia to New York with a banking crisis (1907) in New York that was solved by a syndicate of bankers which acted as the last resort lender under Morgan's leadership. At the beginning of the free trade era (1997–1998), under the Pax Americana, a currency crisis that spread in Asia with a banking crisis was stopped when confidence came back with the IMF action.

The US real-estate bubble burst in 2007 with a great systemic crisis that spread to the world banking system. The great recession of markets and economies showed how quick the contagion to the global economy could be. It does not seem that regulators could change this immediate contagion risk, but reforms and reform effects cannot be forecast, but scenarios can be built. The best way to think about crisis scenarios and reform consequences is to try to give probabilities to possibilities; this process is a much better process than a forecasting process about contagious crisis process. World propagation of news, emotions and risk aversion is now immediate; propagations of financial crisis are as quick or immediate.

Reinhart and Rogoff's book, *This Time Is Different* [ROG 09], about economic and financial crisis history helps us to understand classic crisis features that finally happen or appear; this book shows that crisis perception under approximate factors was seen in many previous crises. The quick propagation, the immediate online crisis propagation, could be a factor to suggest that this time is different. Even if statistics are not available quickly, the news arrives at a time when the scenario of the recent past is not precisely known; it is not helpful to stay rational. In order consider all possible crisis contagions, it is essential to try to not be surprised and to stay more rational than competitors; better asset management returns can be obtained by this philosophy. Thus, contrarian asset managers can sell their equities very early when crisis probability increases and buy their equities very early when mean reversing process probability is high. With this philosophy, some contrarian managers have performed really well during the first half of 2009.

A scenario building process can be summarized with the following thought processes: how many current scenarios extensions and how many historic scenarios brought up-to-date are to be considered? How many extreme scenarios with a crisis process and a convalescence process are there? How many scenarios are to come in the future with a hypothesis of mean reversing process? Average mean reversing of a past period and average process with jumps taken from the past are the possible hypotheses. Some possible oil price shocks added to the previous process create many scenarios.

1.2. Rational expectations theory and the efficient market hypothesis

1.2.1. *Rational expectations hypothesis and geopolitical risks*

To stay rational in a period when global equilibrium is unreliable as Asia increasingly throws its weight into world trade and the global geopolitical game, the good use of forecasts and sound philosophy are needed; although China has hindrances, it is strongly willing to become a global power. China is the Asian regional power on the rise; Asian nationalisms are increasing by reaction. David Moisi, professor at King's College London and adviser at the French Institute of International Relations (*Institut Français des Relations Internationales* (IFRI)), wrote that he saw on the Chinese CCTV4 a message about the Chinese navy in the China Sea being equivalent to the Russian army in Ukraine. Cold war in Asia is a possible scenario even if Mr. Xi received Mr. Abe in Peking. Painful memories are still a major problem and quick Japanese development was humiliating for China which must do better and be much more powerful than Japan.

The medium- to long-term geopolitical competition is impossible to forecast; stable equilibrium is not possible with China competing with the USA in the geopolitical game, economic power and financial dominance; to build many possible scenarios is a way of thinking precisely what is possible for staying rational, to have focused discussions on the most probable scenarios, on the most interesting risks to measure with scenarios including the worst-cases. These kinds of long-term scenarios can have a meaningful effect on medium-term scenarios, but these effects must not be overstated.

Capitalism is based on confidence in the future and on innovations, when trust disappears, and doubts, fears and irrational behaviors are not possible to forecast, but many scenarios are possible. This is valid on geopolitical issues as well as on political or financial issues. The US creativeness and ability to innovate is by far the best; research and development by the US universities, US firms and US foundations sponsoring start-ups are unrivalled. It is not easy to build long-term scenarios on this factor, but it is rational to think like Schumpeter in that it is essential to keep the strong will to win the worldwide innovation competition. On a five-year horizon, to extrapolate the US excellence is rational.

As Communism fell, the common foe disappeared,; therefore, a quasi-cold war between the Islamic world and the West is spreading, and this could justify many mid-term war scenarios in the Middle East and a dangerous security crisis in Europe. Western endeavors to make democracy and western values universal are not acceptable for Islamists: *gharbzadegi* that means *occidentoxication* as Samuel Huntington puts, it is a factor of Jihad in [HUN 96]; western growing understanding of the Islamic threat is only linked to terrorism, nuclear proliferation and dangerous migrations. Resentment against former colonialism, against western military and nuclear power rejuvenate historic rivalry on religious and political conflicts; so, *Jihad* is spreading against a Godless Occident in decay. Much unsteadiness is also likely in Eastern Europe. A new cold war with a Russia who is willing to recover the Russian Empire is a possible worst-case scenario.

Quasi-cold war scenarios could include a progressive Russian conquest of the lost European provinces of the Russian Empire (or the URSS empire); about this, nothing could be forecast; however, many scenarios are possible; some of our learning comes from the Ukraine war and previous conflicts (Crimea and Georgia); to stay rational when there are so many possible scenarios is not easy. But the Baltic States are training rapid reaction forces, for example Lithuania which is traversed by Russian trains, even military trains going from Moscow to Kaliningrad; it could be easy for the Russian army to cautiously leave some intelligence service men and some other soldiers without uniforms along the track in Lithuania. In the 21st Century, uncertainty is rapidly increasing; so a sound rational philosophy is needed to understand the rational expectation's hypothesis validity field.

1.2.2. *Imperfect knowledge and forecasts imply surprises*

Frydman and Goldberg's imperfect knowledge economics (IKE) showed [FRY 12] how a probabilistic model can explain forecast revisions of asset managers; but in major crises, forecast revisions generally occur after the beginning of the crisis in spite of increasing leverages linked to dangerous euphoria, and in spite of discernible credit standard lowering.

Friedrich Hayek and Milton Friedman wrote [FRI 12] that "economic rationale for constraining officials to follow fully predetermined rules stems from inherently imperfect knowledge about how economic policies affect individual behavior and aggregate outcomes over time".

Examining central banks' non-conventional programs, which began in 2007, it was not possible to anticipate them and also their surprisingly successful results, but scenarios could have been built by the Federal Reserve System (FED) observers; giving them probabilities was most uncertain. So it is absolutely necessary to recognize information limits, forecasting limits and knowledge limits as central bankers have to give a limited knowledge of their policy.

Central bankers are surprising when they innovate: for example, when lending of last resort concept asserted itself, in 1826, as a debt crisis erupted in London with many banks runs in London, many crowds in the streets were queuing to withdraw their money. Without any text authorizing the Bank of England to be the lender of last resort and without big banknote reserves, the governor decided to create this function of lender of last resort; the next morning, the bank runs were over. The same kind of scenario has been seen several times. For instance:

– in 1848, with the Bank of France;

– in 1908, with a syndicate of bankers chaired by J.P. Morgan;

– in July 2012, with Mario Draghi speaking from London. He was able to find the right words to hearten the markets, without an explicit guarantee to be a lender of last resort.

The surprise was that he was disregarding Bundesbank's and the court's attitude; Mario Draghi's quasi-guarantee speech in London put German opposition in a corner: to oppose, to censure Mario Draghi's definition of a European Central Bank (ECB)'s guarantee of lending of last resort would imply the end of Euro. Thus, the Karlsruhe supreme court finally decided to transmit that kind of decision to Brussels which supported the Euro. The bond markets of southern Europe moved back to normal levels. So, central bankers are surprising as they hide more than they disclose.

The ECB, lender of last resort, takes an important risk because the natural rate of Germany and north-European states is significantly lower than the natural rate of southern Europe. When a central bank sets a short-term rate lower than the natural rate, cumulative processes of expansion will come with bubble expansions; a short-term rate higher than the natural rate is a brake for economic activity and could result in a cumulative process of recession. A new process of crisis could be seen with a very costly lending of last resort:

– some abnormally low rates and low volatility patterns could be seen in southern country markets, but it could imply very strong volatility ahead, some technical analysts can predict these situations on a short-term basis. World instability is not helpful for forecasters;

– world propagation of news, emotions and risk aversion is immediate; propagation of financial crisis is as quick;

– economic and financial history helps us to understand what happens; Reinhart and Rogoff [ROG 09] showed that crisis perception underestimates factors seen in many previous crises. The quick propagation could be a factor to suggest that this time is different;

– as statistics are not readily available, the news arrives at a time when the scenario of the recent past is not known precisely; it is not helpful to forecast, but building likely scenarios weighted with probabilities is always possible. Scenario generator building is always possible.

1.2.3. *Rational expectations hypothesis and imperfect forecasts in a crisis-prone period*

The rational expectations theory appeared in 1961 with the works of John Muth [MUT 61]; this theory became very popular and had its full development with Robert Lucas's Nobel prize in economics (1995). Wicksell, Keynes and Irving Fisher studied the crisis because their most active part of their career was in a crisis-prone period. Irving Fisher was at the start of this rational expectation concept, but after statistical studies, he noticed that anticipation capability was medium to long term and very much damped. Wicksell and Keynes rejected the idea that when prices are moving in the same direction as interest rates, there is a rational anticipation capability. They noted the slowness with which banks adapt their anticipations; these topics are discussed by Milton Friedman in [FRI 82, p. 630]. To diminish the uncertainty on monetary policy, Friedman argued for predetermined rules. Thus, Taylor's predetermined rule could help us to diminish irrational behaviors in a world of imperfect knowledge and understanding. Even when Taylor's rule is strictly applied, future prices do not seem to be good forecasts.

Nevertheless, it is well known that future prices are not good forecasts in any market; the learning process approach to rationality does not seem helpful. "The price targeting" rule was Wicksell's recommended monetary policy, more particularly, if some deflationary risk was seen. This policy's results are more uncertain if the central bank is not well aware of market expectations. By now, some of Wicksell's rules for avoiding deflation could become more popular than Taylor's rule.

In Chapter 1 of the book by Frydman and Phelps [FRY 12], two authors are quoted for their papers on expectations model which shows how rational expectations hypothesis foundations can be modeled. "Muth's model provides a particularly simple example of such fully pre-determined models ... such a time-structure in effect invariant assumes that market participants do not change the way they make decisions".

Nobel Prize winner Lucas presumed [LUC 75] that *the right theory of capitalist economies, which arguably thrive on routine change, is a fully predetermined model that assumes that such change is unimportant* in this model; he also presumed that rational expectations hypothesis characterizes how rational individuals forecast future market outcomes.

The "imperfect knowledge economics" (IKE) model was introduced by Frydman and Goldberg in [FRY 07]:

> *IKE enables economists to incorporate both fundamental considerations, on which Rational Expectations Hypothesis*

theorists focus, and the psychological and social considerations that behavioral economists emphasize.

In a classic cyclical pattern of a quiet period, predetermined models are sufficient; when there is a more uncertain period, the IKE model is more realistic.

In the case of a crisis-prone period, it is necessary to explore what is possible, at worst, with a crisis scenario generator in a Bayesian approach to calculate VaR, CVaR and TVaR (probability weighted losses that are bigger than the VaR). It is good to complement this generator which includes irrational behaviors and inefficient markets with a rational behavior scenario generator that takes into account rational expectation hypotheses and efficient market hypotheses.

Economists' forecasts are hazardous: economists' consensus forecasts are more reliable because every economist has good and bad forecasting periods; there are periods when an economist can understand much better than others. As an average, economists' consensuses have a more stable record. Sometimes, some economists are able to understand a large phenomenon much better and much quicker than competitors. For example, Henry Kaufman and Wojnilower were able to give more rational advice to Salomon Brothers customers and Credit Suisse customers than other economists during the crisis-prone period of the 1970s.

Overall, the ability of economists to understand and forecast on a regular basis is questionable: the same could be said for asset managers and other financial operators. Media forecasts are enthusiastic when the markets are up and some write "death of equities" when the market is down. The rational expectation hypothesis validity domain obviously does not include crisis-prone periods in spite of the qualities of Irving Fisher, Keynes and Wicksell's works on crisis. Fear and psychological biases that are usual in a crisis-prone period do not allow a rational translation of information into stock market prices; therefore, the weak form of efficiency is not a valid hypothesis. Fear is a cumulative process as a herd effect up to a mean reversal that could be another herd effect. Wicksell studied the cumulative processes [WIC 98] which are very different from a random walk because autocorrelation is strong in a cumulative process and negligible or absent in a random walk. Experienced asset managers with precise memories of cumulative processes can better anticipate these moves that are surprising for less-skilled managers. Learning from experiences is helpful to be able to maintain rational expectations in a crisis-prone period.

Learning is important at the micro- and macroeconomic level.

For example, Yuliya Richalovska (Economics Institute, Charles University, Prague) wrote in January 2013: "the implications of financial frictions and imperfect knowledge in the estimated Dynamic Stochastic General Equilibrium model of the U.S. economy" with a very interesting conclusion:

> The results of the paper allow drawing several conclusions relevant for Dynamic Stochastic General Equilibrium modeling and policy analysis. In particular, due to the ability to amplify macroeconomic fluctuations, learning can be a suitable framework to simulate financial crisis scenarios and various policy reactions.

Sergey Slobodyan and Raf Wouters (Economics Institute, Charles University, Prague) wrote in November 2009: "Learning in an estimated medium scale DESG model: The implications for the productivity and the monetary shock are very promising: the learning models are able to generate an inflation response to productivity shocks that is very rapid and short lived, while the response to monetary shocks is slow but very persistent".

1.2.4. *How could the* homo economicus *be rational with media advice as the only information?*

For the *homo economicus*, an average citizen whose opinions rely on the media, the rational expectation hypothesis is brave in a crisis-prone period; perhaps, in quiet periods, the *homo economicus* could be rational by using common sense without marketing influence of brokers. Financial intermediaries have economists and strategists, but if it was rewarding to follow them, there would be a huge number of millionaires. The rationality of these economists is difficult to assess because they (and other members of broker research teams) have to produce a large flow of orders to keep the brokerage house in good financial health. Marketing and rational advice is not frequently coherent.

Asset managers and other operators have to gauge the risk of every piece of advice and any piece of information. Good asset managers make the right decision 60 times out of 100. An asset manager who is wrong 60 times out of 100 in ordinary periods will not stay long in this position. In a crisis-prone period, there is risk in information, risk in all advice and risk in all forecasts. The risk of irrationality in the synthesis made by an asset manager when he/she is very fearful and upset by a crisis is large. In a crisis-prone period, the rational expectation hypothesis is brave.

This neoclassic theory of an indispensable hypothesis is very handy because rational economic agents, being as rational as economists, give an equilibrium that is easy to formulate for economists; but a convenient market equilibrium with risk equilibrium based on rational choices requires relevant information and psychologies without biases. In a quiet period, such as the last 20 years of the previous century, with mild crises and classic cycles, this hypothesis can be used even if perfect knowledge and perfect understanding of relevant information is not possible.

1.3. Irrational crisis behaviors make previous expectation hypotheses dangerous

1.3.1. *Irrational crisis behavior and fear*

Deflation and high inflation processes result in crises and cause fear through irrational behaviors. In turbulent periods, in a crisis-prone period, behavioral biases and clearly irrational behaviors add risk to intrinsic uncertainty and exogenous uncertainty. Without fearful environments, rational behaviors after mistakes linked to imperfect forecasts and corrections can be modeled; it is different for irrational behaviors linked to fear. Modelization by autoregressive processes with routine shocks and non-routine shocks taking into account the behavioral bias is complicated and could be mind-boggling if fear leads to blunt risk aversion. Without fear, a correction process with a model learning setting behaving correctly is possible. In a normal cycle recession, there are few irrational behaviors which are not based on fear.

1.3.2. *Irrational crisis behavior and bubbles*

Importantly, when a bubble bursts, there are a lot of equity sales, at the bottom of bear markets caused by alarmist papers, journalists and brokers. Some newspapers (for instance, [BUS 79] and [BUS 12]) wrote on the front page "the death of equities" or the "death of markets" right at the end of some normal bearish markets. This kind of newspaper's front page helps them to sell a lot of papers.

Irrational behaviors are seen in the bubble's euphoria and at the top of bull markets when there is a widespread euphoria. When there are important irrational anticipations, they have to be considered as exogenous facts; before the widespread acceptance of a rational expectation hypothesis, many economists used to see anticipations as exogenous facts.

1.3.3. Irrational crisis behaviors and mimesis

Some people, even some economists think that bubble followers as a trend could be rational; it seems difficult to admit that the herd effect is rational as information on herd followers is not relevant but characterized by an asymmetric influence of oligopolistic information brokers. Herd effects and mimesis are important in Korean, Japanese and Taiwanese markets, but not in the Chinese market as it appeared in many papers [CHR 95, CHA 00, HWA 04]; these papers are quoted by Hassari and Rajhi [HAS 14] who found that in Europe, mimesis is moderate in western Europe except in Ireland without mimesis and the Netherlands without significant mimesis. Eastern countries have the kind of mimesis of emerging countries.

Algorithmic trading based on technical analysis and some other kinds of signals is growing quickly. This leads to more herd effects, particularly in developed countries' equities markets as shown by Sornette and Von des Becke in [SOR 11]. In addition to this kind of algorithm, short-termism, which is usually traders' methodology but even many institutional asset managers' approach, increases mimesis.

Classic technological or real-estate bubbles are seen roughly every 10 years in the USA; some unusual bubbles linked to fashionable excessive marketing catchphrases could burst with a crash. All fashions lead to anti-fashion. The end of the exchange-traded fund (ETF) fashion could be the burst of a huge bubble, leading to crashes and irrationalities.

1.3.4. Irrational crisis behavior and illiquid markets with mimesis and some fears

When there is an abnormal recession or a crisis which results in a mix of mimesis and widespread fear, it is obvious that many agents become irrational. Volume disappears and some markets become illiquid because mesmerized operators are unwilling to trade. Therefore, rational hypothesis is then simply incorrect. In a crisis process, information is insufficient and could be misleading; markets cannot be efficient. In illiquid markets, the arbitrage-free concept is a joke. The arbitrage approach of equilibrium is unwarranted, in somewhat liquid markets, as a crisis process unfolds; in this kind of process, non-totally illiquid markets are volatile, shallow and noisy as there is no steadying and logical co-ordination. Expectations are mostly irrational and exogenous in this stressed environment of mimesis and spreading fears.

It is well known that traders and managers take more risks when they have recently won and less risk when they have recently lost. Is it rational? Would it be better to be contrarian? It is hard to be contrarian betting against mimesis.

Cambridge neuroscience researchers Coates and Herbert [COA 08] published a paper on the biological explanation of these behaviors; the excessive behaviors resulting from excessive growth or decrease of risk taken by operators lead to irrational behaviors and ego reactions. In crisis periods, economist's forecast consensuses are unreliable, and risk taking, on average, cannot be more reliable: risk expectations based on liquidity risk, flow of funds, momentum, solvency in recessions or in crises cannot be reliable. In crisis periods, risk aggregation is unreliable because the variance-covariance matrix is unstable. Households influenced by the media cannot be, on average, better than professional risk takers, especially during a crisis. Fear propagation by the media is hard to avoid; mimesis is hard to avoid.

1.3.5. *Market efficiency is not easy to study during a crisis*

Many statistical studies on random walk and efficient markets were made in the 20th Century. Stock market organization changed recently: multilateral trading facilities were authorized in 2007. These new facilities, generally bank subsidiaries, are strong competitors to traditional trading facilities because they built an automated order process capable of trading on many markets without human intervention. High frequency trading and algorithmic trading using market signals to trigger trading strategies (such as high frequency short-term orders of trend following) became usual in trading accounts. Nowadays, many banks manage this sort of strategy in their own right. These high-frequency trading operations produce more than half of the

US markets' orders and more than a third in European markets. So, the multilateral trading facilities' market share is increasing, even if the number of executed orders is much lower than the number of orders sent to the market by these facilities.

Is the liquidity of these markets increased by these new facilities? It is debatable in an ordinary period, but surely not in a crisis period. Volatility is increased and excessive reactions on the news are frequent. Many crashes were seen to be linked, or not, to technical failures. Some are difficult to analyze and understand. All this seems difficult to follow and regulate. When will sufficient time have elapsed with this new organization? Studies on random walk applied to these markets compared to studies made for previous periods will be helpful to understand these market transformations. Are these newly modified markets more efficient? Black pools' lack of transparency do not help.

These stock markets transformation effects cannot be studied in the 2007–2009 crises.

The main drivers of this great recession were the real-estate bubble burst and the credit bubble burst. The efficiency of all markets was disturbed by media's inaccurate comments; bond markets and some other rate markets became illiquid later than the money market. It is very difficult to study stock market efficiency in this crisis. Media information was too complacent in 2007 and too fearful in 2009; individuals able to stay very rational are not very numerous. So, the cumulative process of fear was difficult to reverse. Fortunately, there were contrarians, the kind of courageous and experienced managers who bet on the success of monetary and budgetary reflation policy. This bet on the US economy's convalescence launched a great bull market for more than 5 years. Such a bull market very much helped the economy as it is a great help for confidence in the future:

– Keynes did not build, for his general theory [KEY 36], a general model; partial analysis needed by the heterogeneous nature of the economy and the finance world was made with a pragmatic philosophy of uncertainty and rationality in front of uncertainty. In Keynes' philosophy, rationality is rather a reasonable pragmatic synthesis in front of the pervasive uncertainty of economic and social life. It is a reasonable pluralist approach which takes into account social life and conventions and the lack of information on heterogeneous economic behaviors that could be more or less reasonable or irrational but mostly irrational in a crisis. The lack of reliable information and the behavioral bias in crisis periods cannot mean that the market efficiency hypothesis is verified, not even a weak form of this hypothesis.

– Wicksell's cumulative process from are temporary equilibrium to another was driven by rational anticipations at the beginning of the process, and by irrational anticipations and the herd later on. Wicksell's theory and modeling of value and capital pointed out rational valuations. The mean reversing process with irrational behaviors becoming less frequent must be better analyzed. These Wicksellian studies of cyclical mechanisms [WIC 98], which could be updated with a scenario generator philosophy, lead to cumulative processes in stock market prices with strong autocorrelation (and not random walk without autocorrelation).

1.3.6. *Economic scenario generators can take care of rational and irrational behaviors with some fears and mimesis in a crisis-prone period*

Using many scenarios, generators are able to take into account all these heterogeneous partial analyses (with kinds of rational scenarios and irrational herd following and other mimesis scenarios); each scenario is a partial modeling so that the whole uncertainty which surrounds the economic, social and political life will be measured by the VaR and other risk measurements. It is possible to measure risk excluding irrational scenarios and to compare it to global measures; this could be useful to see, particularly the effect of irrationality on risk measures and to bear in mind the weight of rationality overall.

Many crisis scenarios' mean reversing processes could be taken into account; some are to be built on influences by the size of the risk aversion move and other kinds of fear effects. Risk aversion instability could be a major problem showing an unstable fear process that could very much lengthen the crisis; this could be a contingent rationality process. The number of possible scenarios can be very much increased: each scenario has a derived scenario with a strong effect of irrational behaviors and mimesis so that mean reversing will be late; but another kind of derived scenario with less effects of irrationality and a quicker mean reversing process is justified. Perhaps, another derived scenario with still less effects of irrationality and an even quicker mean reversing process is perhaps justified. The probability of these scenarios is difficult to evaluate, taking into account the past crisis, but it is the only way to take into account various kinds of irrationality and learning processes that could limit or increase the irrationality effects on measures. Generators can include:

– a forward risk-oriented philosophy to cope with black swans with major crisis and illiquidity scenarios, unstable volumes, premium liquidity risks and correlations which cause uncertain risk aggregation;

– a backward philosophy which includes historic scenarios and modifications on these scenarios. To globally measure the risk of a fund, net asset price moment estimating on historic scenarios is not a solution, but to estimate net asset price of a fund for other kinds of scenarios is not easy; to measure market risk, index value moment on historic scenarios is not a good solution; to evaluate index levels for other kinds of scenarios is not too difficult; but which moment? Downside risk measures and maximum drawdown are more helpful if they are estimated on scenarios. By far, the most usual risk measure applied to stock prices is variance, even if in bear markets shallow markets cause a lot of noise which can add up, being averaged, to the half of variance. These noises are dwarfed in the kurtosis, as the fourth order implies that the small moves add up to very small effects on the total kurtosis level. These noises are negligible for skewness being of order 3. As a measure of risk and asymmetry, skewness is interesting for building a more sensitive portfolio to move up than to move down.

When volatilities are unstable, return distributions are fat-tailed. When there is a crisis process as in the 21st Century, distribution can be very fat-tailed. Extreme scenarios lead to illiquidity, fat tails and very uncertain measures of risk and returns; even risk-free rates are uncertain. Expected shortfall measures could be useful with fat-tailed distribution such as VaR, CVaR and TVaR; calculated on past figures, it is an understatement because risk has increased since the beginning of the 21st Century; worst-case scenarios have to be built on future events (in particular, black swans). An economic scenario generator devoted to a crisis-prone period with scenario probabilities given by experts gives the opportunity to evaluate, with realism, VaR, CVaR and TVaR. Economic and financial scenario generators are the right method for including all these complexities, aggregating many kinds of scenarios to obtain meaningful measures of risk, and any kind of measure of risk.

To aggregate risks, there are some statistical problems: correlation coefficients and the variance-covariance matrix are very unstable in a crisis process and consequently, these tools are unreliable in the aggregation process. Copulas could be better, but the use of Monte Carlo methods based on scenarios that come from a generator with copulas is the best.

1.4. How large is the rational hypothesis validity field?

1.4.1. *US mutual fund record and rational hypothesis validity field*

What could be the most efficient market in the world? Without wavering, the answer will be the US equity market. Statistical studies quoted in section 1.3.3 showed that mimesis is not too much of a problem if the real-estate bubble and hi-tech bubble are excluded. Strong contrarian hands, long-term asset managers, and old hands help this US equity market from their example and rationality.

In a published article of February 1999, the 1934, 1937 and 1944 conclusions of Alfred Cowles' works were quoted by Walter [WAL 99]: professional asset management does not show significant added value. This can be proof that efficient markets were difficult to beat from 1903 to 1928. It is not a surprise to see that from 1928 to 1932 results were erratic and difficult to judge, but Keynes who was the economist most capable of understanding the result of the deflationary policy of President Hoover and Treasury Secretary Mellon obtained erratic results (1928–1932) for King's College, Cambridge, that he managed, finally, rather successfully. Does it mean something about the rationality of the US equity market in the great depression? The US mutual funds results were erratic, and finally not as successful as Keynes management. What could a *homo economicus* do to manage his saving in this great depression? Could this average citizen be rational when the media were so shortsighted? It does not seem possible because fear was so widespread.

Michael Jensen's works in 1945–1964 (published in 1968 in the *Journal of Finance* [JEN 68]) concluded that the US mutual fund results under review were less volatile than the market but these fund returns were not significantly better than the market index return. So the return/variance ratios were significantly better than those of the market. It was not too difficult to beat this, mostly, bull market. Could it be seen as a sign of the inefficiencies of this market?

In the short term, Alpha, more particularly Jensen alpha, is positive when the asset manager's stock selection is right even if the market is not so good, as Grinblatt and Titman wrote in the conclusion of their work [GRI 89]. Indeed, they conclude that, finally, the problem with the Jensen measure resulting from timing-related estimation problems can be overcome by a new measure that Jensen called the "positive period weighting measure". Grinblatt and Titman, in subsequent articles [GRI 85] and [GRI 89], concluded that Jensen measures concerning growth funds and aggressive small growth funds showed the capacity of stock selection. Many funds were outperforming (particularly, in their very active periods with a rapid portfolio turnover). The alpha of a portfolio is positive, in the long term, if all losses that occurred – materializing risks during this period – did not go beyond the following term:

(Portfolio return - index return) x beta.

1.4.2. To judge the rational hypothesis validity field is complex

Sharpe showed [SHA 92] that to judge the rational hypothesis validity field is complex. To illustrate this point, he wrote in 1992: "Asset allocation: management style and performance measurement, an asset class model can help make order out of chaos". This study showed that global studies are insufficient for judging whether a market is efficient because some sectors could be less efficient than others. Brown and Goetzmann [BRO 95], quoting Grinblatt and Titman's 1988 and 1992 works, showed that many mutual funds have alphas leading to persistent stock selection capacities. They indicated that funds with "trend following" strategies and "dynamic rebalance strategies" could outperform with persistently good rankings. Many hedge funds used these strategies with successes. After a large survey, it appears that it was possible to outperform the market with persistent and significantly higher returns than the last bull markets of the 20th Century. Does it mean something for the efficiencies of these markets? Grinblatt and Titman [GRI 92] concluded that the mutual funds' past record requires information suitable to judge the ability of the fund manager to outperform in the future. Hendrick *et al.* [HEN 91, HEN 93] concluded that the last two years, and more particularly the last two quarters, provided the most useful information for judging the ability of the fund to outperform in the near-term market.

Brown and Goetzman studied the outperformance persistency variability of funds according to the kind of period and the kind of market; the fund of funds buying funds after a careful study of their record have cyclical risks. As a consequence, a crisis could be a major problem. Furthermore, the 21st Century crises showed that funds of fund are risky in an abnormal crisis. On the contrary, an ordinary cycle gives the opportunity to profit from some ranking persistence: the 20% best in fund ranking for the last two quarters and the last two years would be the best choice. Hendricks concluded that the quarterly fund selection method has to be done on the best 8% in the fund ranking [HEN 93]. Studies on alpha showed that selection on the alpha ranking has to be done on 3 years; alpha persistence on 1 or 2 years is also useful. Many studies (e.g. [LEH 87, GOE 94] and [ELT 96]) showed alpha persistence. Overall, it is difficult to outperform markets. This means that markets are often efficient, but when the forecasting ability is diminished by exogenous shocks, abnormal recession or a process of crisis, that could be very different. This gives an idea of the validity field of the rational expectation hypothesis in spite of the complexities.

Regulations have been increasing since the Pittsburgh G20. Franck's law [FRA 11] was quickly enacted, but its applications were slow and imperfect; many market authorities are slow moving, lenient or unwilling to enforce completely regulations. The European Securities and Markets Authorities (ESMA) studied and proposed to the European Commission an enactment (2014) to be put in force (January 2017): i.e. Markets in Financial Instruments Directive (MIFID II) (MIFID I was only about equities and MIFID II is an overall regulation) and Markets in Financial Instruments Regulation (MIFIR) directives [EUR 14] which were transposed and modified in each country of the European Union. A regulation to avoid market misuse has to be applied directly on European markets without national transpositions; it is known as MAR (Regulation (EU) No 596/2014 of the European Parliament and of the Council of 16 April 2014 on market abuse (market abuse regulation)) (April 2014). These texts will be discussed as they include many exceptions; lobbyists are at work and these texts could, finally, be without teeth. To regulate computer-based markets with one order out of three, or even out of two, being an over-the-counter transaction, is difficult. To diminish the market share of over-the-counter transactions, it is not easy to diminish over-the-counter computerized order flow without decreasing useful program trading, useful index arbitrage and useful algorithmic trading; a fair and protective regulation for investors is difficult to enact and more difficult to apply efficiently.

We can conclude this section by stating the fact that in a quiet cycle market, market efficiency hypotheses, the capital asset pricing model (CAPM) and the Gaussian low are really helpful tools for modeling the evolution of financial values. Moreover, in this kind of period, these hypotheses are almost correct. On the contrary, all these concepts must be called into question in a crisis-prone period: to consider these assumptions as true can be misleading and can lead to bad results.

1.5. Conclusion

Finally, it is necessary to classify scenarios into categories to decide which categories are to be included and which scenarios are necessary in each category, for example, in the category of scenarios based on historic references whose crisis is to be used as the basis for scenarios different from historic figures from one or two parameters (inflation rates and or growth rates); in the category of forward looking, scenarios based on trends and expert opinions. A diversity of scenarios based on an expert opinion and some parameters are to be selected. Comparisons of these categories and some others based on respective weight of rational and irrational behaviors also give ideas on the link between risk measurement and rationality. Many aids thought can be obtained by the use of a generator. Risk measurements for scenario categories give an idea of the link between risk measurement and rational objectivity. When a crisis begins, a generator change can add much help to rational asset management. This managerial decision has to happen before it is obvious that money market instability or illiquidity means that risk-free asset is no longer risk free. The same could be said when sovereign bonds become risky and/or illiquid. With an economic and financial crisis scenario generator, risk management neglects some irrationalities linked to fear and herd effects.

In a process of crisis, risk aggregation in each category, or globally, comprises a major difficulty. At the beginning of a crisis, correlation coefficients increase quickly; on the way out, there is a correlation coefficient mean reversal. The use of copulas does not give any help; therefore, a judgmental approach is needed to build a risk aggregation process. These global risk evaluations are based on two judgmental approaches: probability assessment for scenarios and a risk aggregation process.

2

Scenario Building Processes

For tactical asset management, short-term convictions are needed; usually, a manager has in mind a one-year scenario and an alternate scenario; Solvency II and own risk and solvency assessment (ORSA) set scenario building and risk projection on a horizon of five years maximum (three worst-case scenarios are needed for life insurance companies); in difficult periods with bad visibility, it is necessary for asset managers to study many scenarios and to keep in mind, for example, the four most probable scenarios; in a crisis, it is good to have several or many more scenarios with various kinds of crises and various kinds of ways out and convalescence.

For an asset manager, to be too short-term minded is often a problem; it is good to have a strategic view; however, for a 10-year horizon, so many scenarios need to be studied that it is not realistic to work over such a long period: a five-year horizon is useful and convenient. It is possible to have precise views and beliefs; it is much easier to think precisely over five years than over 10 years. Over a five-year period, decisions and bets result in fewer mistakes than over 10 years. That could be an explanation for the Solvency II horizon over five years, even if insurers are long-term players.

If we take a normal cycle (usually 4–10 years long) as an example, a five-year scenario would cover a cycle or half a cycle; over a one-year scenario, there will be the next stage of the cycle or not. Over a long period without historic crisis, return volatility increases, nearly, as the square root of time has elapsed since the beginning of the period; with a five-year economic scenario generator, it is possible to calculate risk measures; the same computation can be done with a one-year scenario generator. This gives two points of the return volatility curve, which gives a precise definition of the kind of asset management being done.

The 21st Century seems to have analogies to other crisis-prone periods (the 1930s, the Industrial Revolution and the British-dominated globalization of the beginning of the 20th Century). We are in a rising risk century, which could be a geopolitical crisis-prone period resulting in many kinds of crises; a one-year scenario can include a crisis beginning or not; a five-year scenario can describe the full report of a crisis or a partial crisis process.

The number of possible scenarios is much larger than it was at the end of the 20th Century; therefore, to think about all these scenarios and to take account of them with an assigned scenario probability is such a time-consuming job that it is not possible for a manager to do it, except during his holidays. Economic scenario generators can do it and calculate all measures of risk, carry out a portfolio optimization with various methods, optimize with various utility functions and compare to give extremely useful insights. Reverse optimization is also possible with various kinds of software to give other kinds of insights. A computer can do all this as frequently and as needed for the manager wanting to think through his portfolio arbitrage as deeply, as frequently, and as thoroughly as possible. This is particularly needed in difficult periods when visibility is poor and in crisis processes. New regulations such as Solvency II can be an inducement to use that kind of thorough process in the insurance industry.

2.1. Most asset managers have only one or two scenarios in mind

Different behaviors have been observed by analyzing the market. However, most asset managers have in mind only one or even two scenarios. In the sections below, we summarize the different types of attitude encountered.

Usually, an asset manager has in mind a preferred scenario coming from a firm belief on which bets, even big ones, will be made. If this preferred scenario is bold, an alternative scenario could be taken as a more prudent view, not as far from the market-implied views as the preferred scenario. This alternative scenario being less preferred, it would result in fewer and smaller bets. That kind of situation could be possible when the asset manager is a contrarian betting that the market strong moves are often wrong.

However, this contrarian cannot know when the market mistake will be understood and corrected; therefore, an alternative scenario could be that the mistaken move will go on for some time and some short-term bets could be made over this term.

Another kind of situation could be that the preferred scenario is not very bold, and not very different from market-implied views; an alternative bolder scenario could be used to make some rather small bets.

On the whole, most asset managers do not have more than two scenarios in mind: one firm belief and an alternative. An economic and financial scenario generator can follow a great number of risks and scenarios to be able to calculate many risk measures that cannot be in the mind of asset managers who only work with two scenarios in mind.

That kind of generator can compute risk measures and expected returns taking into account many scenarios: some scenarios can be compatible with market-implied views, and many will be very different.

Another method can help the asset manager who gives probabilities to his own preferred scenario and the alternative scenario, leaving an important probability to other scenarios. To measure risks and expected returns and to compare with results from the generator not taking the asset manager's preferred scenarios could be very helpful. The same kind of comparison could be done with the results obtained with the generator including or not scenarios which are compatible with market-implied views. Another comparison can be helpful in a crisis period; the same kind of comparison could be done with the results obtained with the generator including or not scenarios which are rational and those which are too fearful to be rational. These comparisons are able to help understand the asset manager boldness compared to market boldness. It could be very rewarding especially in difficult periods. A review of the kind of scenarios not taken into account, ordinarily by managers, is also interesting.

2.2. Long-term scenarios and geopolitical surprises

Prices of equities have a more logical behavior record in the medium to long term than in the short term. Economic forecasts over the long term generally have a low probability of occurrence. So, low-probability scenarios, which are not taken into account, must ordinarily be included in the value at risk.

Scientific and technical progress was seen by Schumpeter as a process of replacement of obsolescent products because innovative products are found; this process keeps up long-term growth. Paul Romer's (Stanford University) "Endogenous technical change" (1990) model was based on a growth that innovated with a greater variety of goods and services as the information society grew.

Information and learning progress is such a well-established trend that long-term forecast can be extrapolation. Immediate information is worldwide and learning is spreading through the Internet; information quantity is a problem that could be less difficult and less costly to address with cloud computing for big data and supercomputer networks.

What could be envisaged is that data processing informatics and information has invaded our society in stages. A first data processing bubble was the stock market reaction to the beginning of the computer era in the late 1950s; a second bubble came in the bear market of the late 1970s; this hi-tech bubble up to 1982 was driven by Silicon Valley stocks as microelectronics, microcomputers and even some supercomputers (Cray). The third step was the Internet bubble up to 2001. A cloud computing and big data bubble could be seen in the next five years. This trend could go on bigger and cheaper storage warehouses, more pervasive supercomputer networking, etc.

Most government bodies will build big data surveillance systems (climate monitoring already has big supercomputers), as big companies will build huge databases for their marketing. If there are some totalitarian governments, they could connect all this into a Big Brother network.

2.2.1. Climate change scenarios and surprises

Climate shocks will result in significant prices moves. Some long-term investors already have investment processes taking into account environmental criteria (environmental, social and governance (ESG) criteria, see Box 2.1); for example, the carbon market could, in the future, have important effects on many industrial companies' stock price. As it is difficult, even impossible, to forecast climate shocks, it is necessary to think about climate change scenarios. Moreover, taking into account stock price inefficiencies represents a hedge against climatic shock effects on long-term investor portfolios.

Best-in-class investment means investing in companies that are front-runners in meeting ESG criteria in their particular universe, asset class or category. It is also sometimes called *socially responsible investing* (SRI).

Examples of the ESG criteria that companies and investors will consider are:

Environmental: Are renewable energies incorporated within the manufacturing process? Is there prevention from environmental and air pollution? Is the energy used clean and does it consider climate change? Is waste recycled?

Social: Does the company play a role in social development? Are human rights incorporated into business goals? Are industrial safety provisions in place?

Governance: Are corporate management goals transparent? Are shareholder rights protected?

Box 2.1. *ESG criteria*

In case of very costly surprises, which could be climatic or geopolitical, it is good to break the force of the blow using a temporary subsidy of some goods (as done by President Giscard d'Estaing with the right wing Barre government in 1978 to alleviate the oil shock); to optimize that kind of policy, Romer model optimization could help to maintain some growth with new products and services by increasing product and service diversity. The climate shock could be much more harmful than the oil shock; probabilities of very harmful climatic scenarios are increasing as nearly nothing is done to avoid these probability increases.

Climate conferences from Kyoto to Lima are on a very slow process; two scenarios are studied: 40 and 70% gas emissions cut. Each country has to prepare a reducing scale of emissions for noxious gas with commitments by industries. A study by the United Nations made valuations of the necessary investment program ($70 and $100 billion scenarios) to limit climate change consequences; a green fund to help emerging countries to limit these consequences; 27 countries gave public money to this 10 billion fund. Small countries asked for an indemnification fund to be built. Some private sector endeavors add up to these public sector endeavors.

The United Nations housed portfolio decarburization fund (this fund's aim is a $500 billion portfolio with a minimum carbon exposure footprint), which is a multilateral initiative of the United Nations environmental program financed to induce institutions to buy carbon-efficient companies and to sell carbon-intensive firms. Mr. Anderson (Chief Executive Officer (CEO) of Sweden's fourth largest pension fund) received a prize award from "Investment and Pensions Europe" for an index-based strategy (adopted by the United Nations) with companies producing very low greenhouse gas emissions. The aim is to induce company commitments to diminish greenhouse gas emissions. The UK Environment Agency Pension Fund uses carbon-related risk criteria in investment strategy and climate-related risk engagement. This fund supported requirements for corporate disclosure for each type of fossil fuel. These initiatives are taken to begin, modestly, a process of limitation of climate change risks. Martin Wolf and Hank Paulson (former U.S. Treasury Secretary) warned about companies exposed to climate change risks.

Since the early 20th Century, the global temperature has increased approximately 0.8°C with about two-thirds of the increase occurring since 1980 [NAT 11]. Each of the last three decades has successively been warmer than any preceding decade since 1850 [IPC 13] Consequently, this phenomenon has to be taken into account in long-term scenarios. Nevertheless, we have observed that climate change is not much taken into account by asset managers.

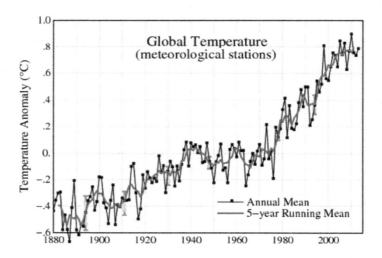

Figure 2.1. *Evolution of the temperature from 1880 to 2012 (source: NASA). For a color version of the figure, see www.iste.co.uk/clément-grandcourt/forecasts.zip*

Some long-term scenarios have to be built on global warming: the average temperature could be 2–6°C higher at the end of the cycle. The worst-case scenario over five years is 0.5°.

The climate change manifests itself as rain and water increases as well as severe drought which could lead to important agricultural production surprises. For example, the table below summarizes several catastrophes that have occurred over the last seven years.

2008	Exceptional temperatures in China and Argentina
2009	Record heat wave in south east
2010	Floods in Pakistan, in Australia; extreme drought, heat in Russia
2011	Extreme drought in western Europe and US corn belt
2012	Extreme drought in the USA, Russia and south east Europe
2013	Floods in Germany, Czech Republic and Canada, hailstorm in France and Germany, hurricane and tornadoes in the USA
2014	Floods in Bosnia and Serbia, tsunami in Chile, floods in Bolivia, hailstorm in France and Germany

Is that a list of surprises? Agricultural raw materials inventories are low now compared to previous periods: since 1945 and up to 2008, fertilizers implied excessive inventories of agricultural raw materials. Consequently, climate variability did not change much the low prices of foodstuffs during this period. However, climate change could have a significant effect in the future. The analysis of the price moves over the period (2008–2012) shows that cereal prices could increase by 50% in case of some weather problems that could arise rather frequently; a 100–200% increase would be a climate problem. Indeed, this kind of increase could result in significant political problems. For instance, the recent upsets and unrest in Tunisia and Egypt are the result of a shift in the price of foodstuffs. It could be compared to the 1789 and 1848 French Revolutions that happened after three years of bad weather and a price rise in cereals at a time when food was an important proportion of French household budgets. There are still populations for which food is the main proportion of the budget. The Tunisian and Egyptian governments tried to avoid social crisis with subsidized foodstuffs as much as they could.

However, we have to keep in mind that there could be a mean reversing process.

2.2.2. *Climate changes and migrations*

Even without the worst-case scenario, variability and all kinds of atmospheric disturbances are already increasing and could accelerate. Casualty insurance and reinsurance are already working with short-term scenarios including increased costly climate casualties, but medium-term big surprises are considered. These scenarios could be integrated into a crisis scenario generator with or without social crisis scenarios induced by food prices. Water and food major problems imply migrations, social upheavals, political uprisings and wars; long-term scenarios of strong climate change could mean geopolitical surprises over the long or medium term.

Huntington's works [HUN 96, HUN 97] show that the geopolitical equilibrium becomes increasingly unreliable because the balance of power between the West and the rest of the world is moving rather quickly, to the prejudice of the West. Migrations could be a major component of this kind of scenario.

2.2.3. *Geopolitical scenarios and surprises*

On the health expenditure and pension fund equilibrium, the health and science industries could mean very significant cost surprises, and social and political actions could involve big surprises too.

In Europe, surprises can come from the interaction of aging and migration trends at any time; in Germany, a 1.5 million downward statistical correction was done on the total number of German inhabitants after taking the census, but an immigration flow of young people with good diplomas is coming from southern Europe, especially Spanish engineers attracted by the German industrial excellence of some industries. How many will go back to their countries as experienced managers able to start new firms in their home countries?

The consequences of aging processes are very important, as well, for China as a future economic, financial, political and geopolitical power or superpower; it involves many scenarios, even over five years. The average Chinese person has an income of 20% of the average Japanese person. Could equality be reached in 2025, 2030, 2040 or 2050? This implies four scenarios for China and scenarios of effects on Europe and on the Eurozone industry that are possible to build schematically. It is obvious that China is willing to be a world power but many social and political problems could occur. Many geopolitical scenarios are possible; the worst case would be a war or a long-term cold war turning to a global conflict in Asia: China, Pakistan allied with Iran, Iraq and Afghanistan against Japan, India and the USA. Wars and the aging process could be linked. Aging is an easy to forecast process for these countries, even, over the long term, with few parameters; scenarios have been built for a great number of countries and used in the economic forecast process.

The USA has the best demographic future among developed countries due to migrants from Asia and Latin America. Japan has the worst, and China is not as bad as many people think. Portugal, Spain and Italy are worse than Germany because there is a new flow of migrants. Many people are coming from southern Europe which does not help these countries to recover their equilibrium.

Middle Eastern wars are scenarios that can be numerous, even over the medium term because the Jihad state is a big upheaval factor which could mean revolution scenarios in many countries with extensive effects on oil markets. History teaches that young people used to fight aging civilizations.

2.2.4. Long-term demographic impact

Populations and migrations forecasts

United Nations 2012 estimates show world population figures are still growing quickly: 7.3 billion in 2015, 7.7 in 2020 and 9.6 in 2050 (US census bureau indicates 9.4). The geopolitical world equilibrium will be changed as Europe has 10% of the world's population in 2015 and (in spite of immigration) will have 7.4% in 2050 as Africa will move from 16 to 25% and Asia from 60 to 54% when northern America edges down from 4.9 to 4.7%. These forecasts imply declining fertility rates worldwide (global fertility rate has been down by a third in 30 years), but it seems that the fertility rate in Africa is no longer going down. India will be the most populous state in 2030 with 1.5 billion inhabitants when China would have 1.4 billion inhabitants.

The United Nations have made migration projections for 40 countries, for a five year period (2005–2050): 98 million migrants that will prevent population decline in 28 developed countries as deaths will be more numerous than births by 73 million in developed countries. Fertility rates are stable at low levels in Germany and somewhat upward in the United States, the United Kingdom and France. By 2050, Germany, France and the United Kingdom will have approximately 80 million people due to immigration. Immigration is not easy to forecast; Germany had 300,000 people arriving after reunification, mostly ethnic Germans coming from eastern countries; immigration went down to 50,000 later. At the beginning of the 21st Century, Italy had a strong immigration that was down by half recently.

European population aging

Life expectancy in most European countries has increased by seven years in 30 years; this aging process is going on unabated; therefore, long-term care cost will increase by 1% gross domestic product (GDP) at least in the next 50 years and healthcare will cost more of the government expenditure budget. Pension expenses paid from the national budget will increase except for Italy, Portugal and Denmark because these countries already have high pension costs; some other countries will have to modify their pension systems. Aging has many more effects on consumption, saving and investment.

The developed world is aging. On the whole, this has negative effects on growth and inflation. However, this also has positive consequences on the growth of some sectors, such as the health sector, real estate and the infrastructure sector (e.g. the US real-estate cycle can be explained by demographic data), financial services, leisure services and even luxury goods. Aging has to be taken into account, for growth and inflation scenarios, with regard to European countries.

2.2.5. *Economic emergence of the African continent*

The African continent as a raw materials producer has benefitted from high prices linked to quick growth in Asia; with a more moderate growth in the Asian continent, commodities prices have more moderate prices, which could mean slower African growth; after the Korean War and the Vietnam War, lower metal prices had a clear effect in African metal-producing countries. Which countries could transform a commodities linked boom in a self-sustaining development? If equatorial countries were not so politically problematic and socially unstable with an abnormal proportion of youngsters and unemployed young people, self-sustaining development would be possible. Development scenarios are more possible for more stable tropical countries.

South Africa and other countries in this area with a growing number of graduates could add to their big mining business, and modern industrial activities. Compared to South Africa, Northern Africa does not have significant commodity wealth; however, there are many graduates who are able to generate some new activities. Generally, Asian development cannot be followed by a foreseeable African development with the same kind of geopolitical consequences on world equilibrium even if demographic development could be more significant than in other continents; it is possible that millions of migrants to Europe could induce some development in Africa over the long term. Two scenarios could be made for Africa: a mean reversing scenario and a moderate development.

There are also two demographic scenarios for Africa based on the fertility hypothesis: fertility rates, on average, could be stable as recently or decrease at the same rate as the rest of the world. Fertility rates have been cut by half in Asia; life expectancy at birth has increased nearly as much as in Europe. Asia is on a different process than Africa: Asian economic development is on a different stage and is a different process than African development that is a much longer term process.

A conclusion about the main long-term trends could be: the climate change calendar is unforeseeable and the demographic calendar is possible to forecast with some migration scenarios. World leadership will become multipolar, but many scenarios are possible. Some people think that China could take the world leadership in a foreseeable future.

People in the USA, who know why 75 years were necessary for New York to take the financial leadership from London and for the dollar to replace the pound sterling, are very sure that scenarios for the yuan being a dangerous rival for the dollar as reserve currency are long-term scenarios. The dollar staying as an unrivalled reserve currency for the short term is the most probable scenario; over the medium term, it is not obvious. Competition between China and the USA could induce instabilities, over the medium term, if the yuan zone is managed in order to destabilize the dollar zone.

The US research will remain very much ahead of the Chinese research over the medium term if the Chinese research development follows a Japanese scenario. The long-term scenario of China overcoming the USA for innovative research and financial leadership has a sizable probability even though China is aging more quickly than the USA which is more able to attract the best brains in the world.

2.2.6. *Long-term risk valuation methods*

Long-term scenarios as the sum of short-term and medium-term scenarios are too simple a concept to be useful in crisis-prone periods. In this kind of period, it is much more realistic to think long term as a process of trials and errors followed by corrections. Apprenticeships on rational and irrational behaviors of economic, financial and political actors induce an ability to judge and cover risks. To evaluate the global risk of these kinds of process can be compared to casualty insurance methods.

Life insurance companies and pension funds have long-term liabilities, so they used to invest over long-term horizon fundamental values. The most probable portfolio value is much more difficult to forecast if their investments are made with successive short-term horizons. Short termism does not give very much weight to fundamental financial analysis research.

The thesis of this book is that scenario studies are helpful to allocate funds to have persistent returns and precise measures of risks; scenarios including mean reversing and different kinds of market error corrections are helpful to judge contrarian strategies that could be costly over the short term and very satisfactory over the medium to long term. To sell in a euphoric period and to buy in a crisis is difficult, but long-term investors are able to do it well; to measure and judge the risks of a contrarian strategy, a scenario generator is useful.

2.3. Five-year scenarios

The tables below summarize different kinds of five-year central scenarios for the world economic situation in terms of inflation and growth evolution.

Evolution of inflation rate:

Country Nb	County name	Number	2016	2017	2018	2019	2020	End of cycle	Long term tendency	Indicator
1	WORLD	111	3,2	3,0	3,2	3,3	3,3	Slight economic downturn in 2016	Optimistic	↑
1	WORLD	112	3,1	2,8	2,9	3,0	3,2	Sharp economic downturn in 2016	Optimistic	↑
1	WORLD	113	3,0	2,0	2,5	2,7	2,9	Risk of recession in 2016	Optimistic	↑
1	WORLD	121	3,0	2,9	3,0	3,2	3,3	Slight economic downturn in 2017	Neutral	=
1	WORLD	122	3,0	2,8	2,5	2,8	3,1	Sharp economic downturn in 2017	Neutral	=
1	WORLD	123	3,0	2,7	1,8	2,0	2,6	Risk of recession in 2017	Neutral	=
1	WORLD	131	2,0	1,8	1,6	1,0	1,2	Slight economic downturn in 2018	Stagnation	→
1	WORLD	132	2,0	1,8	1,4	0,5	0,7	Sharp economic downturn in 2018	Stagnation	→
1	WORLD	133	2,0	1,7	1,0	-1,0	-2,0	Risk of recession in 2018	Stagnation	→
1	WORLD	141	1,5	1,4	1,3	1,1	1,0	Slight economic downturn in 2019	Deflation	↓
1	WORLD	142	1,5	1,4	1,2	0,5	0,0	Sharp economic downturn in 2019	Deflation	↓
1	WORLD	143	1,0	-0,5	-1,0	-2,0	-3,0	Risk of recession in 2019	Deflation	↓

Evolution of growth rate:

Country Nb	County name	Number	2016	2017	2018	2019	2020	End of cycle	Long term tendency	Indicator
1	WORLD	111	2,8	3,0	3,3	3,6	3,9	Slight economic downturn in 2016	Optimistic	↑
1	WORLD	112	1,0	1,5	2,5	3,0	3,6	Sharp economic downturn in 2016	Optimistic	↑
1	WORLD	113	-1,0	1,0	2,0	2,4	3,5	Risk of recession in 2016	Optimistic	↑
1	WORLD	121	3,7	2,3	2,6	2,9	3,3	Slight economic downturn in 2017	Neutral	=
1	WORLD	122	3,7	1,0	2,0	2,5	3,5	Sharp economic downturn in 2017	Neutral	=
1	WORLD	123	3,7	-1,0	1,0	2,0	3,0	Risk of recession in 2017	Neutral	=
1	WORLD	131	3,0	3,2	2,3	2,6	2,9	Slight economic downturn in 2018	Stagnation	→
1	WORLD	132	3,0	3,3	1,0	2,0	3,0	Sharp economic downturn in 2018	Stagnation	→
1	WORLD	133	3,0	3,5	-1,0	1,0	2,0	Risk of recession in 2018	Stagnation	→
1	WORLD	141	2,5	2,6	2,8	2,4	3,0	Slight economic downturn in 2019	Deflation	↓
1	WORLD	142	2,5	3,0	3,3	1,0	2,0	Sharp economic downturn in 2019	Deflation	↓
1	WORLD	143	2,5	3,3	3,5	-1,0	1,0	Risk of recession in 2019	Deflation	↓

2.3.1. *Many five-year crisis scenarios are possible in Europe*

Camdessus voiced [CAM 14] many times that it is sad that to make true reforms, a very important crisis process is needed. Many historic examples could be found: fear is needed to overcome vested interests and political blockades. As a Latin country with a confrontational culture, fearful social upheavals are possible, even foreseeable when people will no longer be afraid for their jobs; the same could be seen in Portugal but it would be a smaller problem for Europe. It is very unsure that Mediterranean countries could by long suffering austerity and deflation erase external balance deficits and wage divergence; the end of the Euro could come with very important sociopolitical events, after all the palliative measures are sold out, i.e. exhausted. The end of the Euro would put many countries in the kind of strong inflationary process that can be found in Latin American countries.

Even if Draghi pledged [DRA 12], in July 2012, to fight to keep the Euro alive because the European Central Bank (ECB) would be a buyer of last resort to avoid Eurozone fragmentation, the ECB cannot be a true lender of last resort; Eurozone defense organisms do not have the clout to stop a hedge fund attack that would propagate a systematic crisis by derivative markets of an unforeseen size. Greek bond market haircuts did not lead to bank failures; is there a probability for a scenario of crisis propagation based on toxic papers and for a scenario of banking failure expectations which would involve a systemic crisis with panic runs on the European banks?

Speculative asset management could be so quick to bet that banking failures could spread to southern European weak banks that authorities could not be quick enough to fight systemic crisis in Spain, for example. The Spanish banking system which did not book all real-estate losses still seems shaky; some banks could fail in the next five years; after a €40 billion recapitalization of the banking system and a lot of bad debts transferred to the SAREB defeasance structure, the Spanish banking crisis is not over. It could induce sufficient signals for hedge funds to speculate against this banking sector crisis propagation. Spain's Mediterranean provinces are deeply indebted and unwilling to pay increased taxes; they are asking for help from Madrid; if Spain receives credit from the International Monetary Fund (IMF) or other official sources, it will have to retrocede an important proportion to the provinces. The Catalonia referendum was forbidden by the Supreme Court as unlawful, but a kind of quasi-referendum was held by regional authorities to try to obtain the same increased fiscal privileges as Scotland. A leftist new party "Podemos" (the Spanish for "we can") has 30% votes in Gallup polls, which is slightly below the two main parties; therefore, the December 2015 general elections could be a problem even if the economy is improved.

The frailty of the Greek political and social situation is also a risk.

There could be isolated crisis scenarios and also a process of multiple crises; the end of the Euro would be a multiple-crisis process: national and international authorities would fight to try to avoid the Euro falling off a cliff, this could be studied with game theory in imperfect information with the Tirole methodology; scenarios of spreading social crisis with major inflation in southern Europe could result from the end of the Euro; scenarios can be built by comparison with some Latin American failures. This could be avoided by a lengthy process of big reforms made because Europeans would have understood what the end of the Euro means; fear of the end of the Euro could be a strong incentive if political leaders were sufficiently courageous to explain this danger.

There are isolated crisis scenarios, and also processes of multiple crises; the end of the Euro would be a multiple-crisis process with social crisis and with major inflation in southern Europe. This lengthy fearful process could imply big reforms made in fear.

In five years, a great number of scenarios can be seen; these scenarios can be classical or not, they can include many kinds of fear propagation, many kinds of irrational behavior and some learning processes. All this can be summarized with judgmental risk aversion scenarios.

Bundesbank Weidman said in [WEI 14] that the Euro was intended to federate European countries to obtain a power able to be a counterbalance to Asian emergence. Euro structural deficiencies have led to a dramatic divergence between northern and southern Europe. Hans Werner Sinn [SIN 11], the well-known professor of economics and finance teaching in Munich, criticized German banks which lent large amounts to the southern countries of Europe thinking to obtain better rates of return; so credit to German small- and medium-sized enterprises has been insufficient for optimum growth. Private sector credit has been diminishing since 2002 with a negative effect on investment. Some German economists, even a government adviser, think that market share losses could be seen by the German industry in the future because investment endeavor has been insufficient.

In the following sections, we describe several possible scenarios for the Eurozone:

– How different could the next Euro crisis be?

– Is Brussels monitoring better?

– End of the Euro?

– What could the next crisis process be?

2.3.1.1. *How different could the next Euro crisis be?*

To build Euro crisis scenarios, it is necessary to wonder what could happen as in 2010, what could be different, how early this kind of crisis could be foreseen, i.e. forecast? Since 2005, academic papers, seminars and conferences have described a divergence problem in the Euro area. This monetary area did not fit with the rules established by Mundell (Professor at Columbia University) for monetary area persistency. The Maastricht convergence criteria emphasized nominal convergence much more than real convergence. Public sector stability is taken into account by these criteria, but not the private sector stability. The great recession showed how unstable the private sector can be. Institutions that are too big to fail can cost a lot for the public sector which act as a lender of last resort.

When Spain's real estate bubble burst in 2007 the state was the lender of last resort. Ireland had a low debt to gross national product (GNP) ratio up to the time when the state acted as the lender of last resort for its banking sector; this state then became one of the most indebted countries in Europe. Interest rate spreads on sovereign bonds are sovereign stress indicators which are dangerously correlated to banking sector stress indicators. Endeavors are being made to reduce this correlation; therefore, this risk could be less important in the next Euro crisis. Three scenarios are needed:

– this dangerous correlation is unchanged;

– this dangerous correlation is reduced;

– this dangerous correlation is no longer important.

2.3.1.2. *Is Brussels monitoring better?*

With the Euro, finance ministers are no longer worrying about their exchange rate currency but they have to monitor sovereign bond ratings and spreads (when there are approximately 100 basis points). 500 basis point spreads mean financial dependence on international authorities. It is simple and clear-cut, so Brussels could do it easily.

Brussels did not react early enough to indicators of divergence about industrial production; market share losses of Mediterranean countries' industries (French industry included). Postcrisis deleveraging is not a help in stopping this process; Brussels did not act much with structural funds. To reindustrialize southern Europe, many reforms are needed; a unified labor market, a unified capital market and a federal budget to finance in size structural funds would help. All of this is difficult to realize.

2.3.1.3. *The End of the Euro?*

Exiting the Euro area is not impossible but it would be dramatic for all countries except Germany; the cost for Germany would be very high but Germany has reserves to recapitalize all failures. Other Euro countries would have the kinds of big devaluation that some Latin American countries experienced. In this case, social problems add to legal, financial and technical problems. These countries would have to drastically reduce imports, especially from Germany. So, it could be less costly for Germany to subsidize southern Euro area countries than break up the monetary union. The end of Euro is, nevertheless, possible as "Alternative für Deutschland" grows. The other scenarios are more likely.

2.3.1.4. *What could the next crisis process be?*

In the case of social and/or political upheaval, the Nordic states and Germany will not see their help given to southern states as an investment but probable grants to states that will be unable to repay because their policies are always lax. Convergence of southern culture to northern culture does not seem to be probable or possible. Sweden with the Wicksellian culture of the central bank was quick to clean the credit bubble which burst in the early years of the 1990s; recovery was quick. It is the contrary to the Japanese deflation mess. Slow and late banking sector healing and insufficient Keynesian stimulus failed to end the deflation; credit minimization obsession and skepticism is difficult to change. The Euro area is strengthening the banking sector but credit demand is weak, and investment is weak because entrepreneurs are skeptical and waiting for recovery to invest; deflation induces consumers to wait for lower prices to buy. A Japanese scenario for the Euro area is not impossible, but the Japanese experience will help the ECB to avoid mistakes. Wicksellian advise: weak currency, moderate austerity except when recession is coming will be a possible scenario; Keynesian reflation will be seen from time to time; some scenarios can be based on these concepts; the most probable scenario cannot be effective very quickly due to legal problems with Germans, fear of moral hazard in some countries and the simple fact that it is hard to convince 18 countries; fear of a mess in case of breakup and sense of survival have been a help in the Euro crisis; it will be a help in the next crisis. A long term refinancing operation (LTRO) can be performed again and again quickly when needed.

2.3.1.5. *A stronger Eurozone is also a possible alternative*

As mentioned above, true reforms are often implemented in a context of crises. Indeed, when we have our backs to the wall, we do not have any other choice but to set up reforms. Consequently, in the current climate, changes can be expected and a stronger Eurozone can arise. For that, various obstacles must be overcome: the opposition between the German vision and that of the countries of southern Europe; the problem of national debt; the

Greek crises, probably followed by the Spanish and the Portugal crises, and so and so on.

In spite of the several obstacles to be overcome, this alternative is possible and a five-year scenario integrating a stronger Eurozone has to be considered.

2.3.2. Some kinds of Japanese deflationary processes

The 1997–1998 Asian countries crisis was an example of a bubble bursting with a currency crisis beginning with Thailand, Malaysia and the Philippines; this crisis became a banking crisis spreading recession. The Japanese global bubble burst as the monetary policy tightened. This global bubble burst with collapses in stock prices and real-estate prices; this pushed Japan into a deflation process with a private debt liquidation process which can be summarized as replacement of private debt by public debt; in 1999, debt liquidation began as well as credit decrease from banks to firms; the global credit amount in 2005 was at the same level as 1996. Quantitative easing carried out by the Bank of Japan from 2001 to 2006 increased the monetary base but not the aggregates. The same problem was seen later in the Eurozone and English-speaking countries.

A cumulative process has been seen: the Japanese private sector was paying down debt; the media made public opinion afraid of deflation. Consumers began to wait to buy durable goods to obtain better prices; distressed sales of goods and stocks pushed prices down; in the beginning, rational expectations were moderately on the downside, but irrational anticipations came when the reflation policy failed twice; the economy trended dangerously on the downside as fear spread. The decrease in prices increased the weight of debt in the balance sheet of firms and the weight of interest payment on costs, so a strong movement of debt liquidation weakened the economy; households had the same problem and tried to deleverage. A cumulative process based on risk aversion induced the enterprises (and even the households) to hoard and as a consequence the drop of the money circulation speed up; so not only did the monetary policy stop working, but fiscal policy was also used to maintain the demand. Since the balance sheet reduction process became strong, the monetary policy became ineffective, and the budgetary policy became crippled by the drop of tax revenues so that private debt was replaced by an increase in public debt. Public issues were crowding out private demand of long-term funds.

Japan have had a full deleveraging process and still have a debt phobia; however, a Keynesian support of the economy and the growth of exports were sufficient to avoid depression but insufficient to escape stagnation; with a bigger Keynesian support and a lower yen obtained by lax monetary policy, economy would rebound. Everybody in the private sector has been paying down debt; debt phobia going too far and too long implies that to reverse this cumulative process is very difficult. So, the mean reversing process is more difficult than forecast. The needed bullish stance is not seen.

With Abenomics [ABE 14], is Japan experiencing a way out? Private debt phobia can last a long time after the end of debt liquidation and could be a brake for growth. With a national debt to GNP ratio of the same size as after war, it will be a long process to get back to a normal level. Cutting public debt is necessarily slow and painful in a quickly aging country; nevertheless, in a country which is not social crisis prone and which is capable of great national endeavors, public debt can be cut.

What kind of probability level could be given to a Japanese scenario depression? When rates are moving up across the world, a prime minister could decide to increase taxes to stop debt increase in order to reduce debt; in this case, a depression could be seen; in a period of diminishing rates, this kind of decision was taken twice with the same result: a recession that was very difficult to fight and to erase; an unavoidable depression could be seen because debt is now so much bigger than the levels seen when previous reflations were tried. Depression implies significant negative rates of inflation that increase the debt impediment.

One-year Japanese scenarios: in a protracted process of weak rebound obtained by money creation and yen devaluation and stagnation, most probable one-year scenarios and weak rebound, one year of deflation is possible; other probable scenarios are stagnation with or without deflation and depression scenario and strong rebound scenario due to aggressive monetary creation or have low probability.

2.3.3. *Different kinds of deflationary processes in Europe*

Using comparisons of Eurozone problems with Japanese scenarios, it seems that weak rebound without deflation is the most probable scenario, deflation is still a threat as the ECB does not have an aggressive stance to fight it or depression; strong rebound would need an unforeseen boom in the USA and/or reacceleration of the Chinese economy which could start an investment cycle in Germany.

Wicksell and the Swedish school showed that in a crisis this is the most important thing to avoid a negative inflation rate with currency devaluation; as English-speaking countries have been creating a lot of money since 2009 to weigh on their currencies, the Bank of Japan with Prime Minister Abe has had to create a lot of money to push down the yen. It is not possible to weaken the Euro as it is needed to help southern European countries out of stagnation for two reasons: German psychology and ECB status.

Wicksell, for example, defined the natural rate in page 192 of his book *Lectures on Political Economy [WIC 01];* he explained how too large a gap between actual rate and natural rate could imply a cumulative process. Eurozone interest rates are too low for southern European countries with excessive budget deficits; consequently, bubbles and excesses of all kinds are very difficult to rectify. Eurozone wage divergence is too big to be corrected by austerity and deflation that could induce possible social crisis scenarios in the south. After an austerity period that the IMF judges too abrupt as the Swedish school would also rate it, a Keynesian reflation plan to boost the economies would be good if there is not the kind of social crisis scenario already quoted: 1936 in France.

Eurozone interest rates are too high for Germany and the northern states of Eurozone; indeed, these levels of rates represent a real impediment to the development of their investments and their medium-term productivity; the Euro has diminished the cost-cutting culture of German industry because its (too low) exchange rate leaves, over the short term, excessive margins to industries. If the Deutsche Mark (DM) reappeared, its price would be trending up; this would induce the emergence of a new cost-cutting culture. Interest rates would be down inducing stronger investment, higher productivity and a better growth later on.

Germany is the only country which can afford its way out the Euro without disruption because her huge positive balance at ECB and elsewhere can be used to recapitalize banks and other firms that could be wounded by the increased value of the new Deutsche Mark, which could be stopped by the Bundesbank at 20% with sales of a newly created DM as the Swiss national bank stopped the Swiss Franc rise at 20%: this means that over five years, there is a possibility of a DM. The lesson from East Germany is: the more you give, the more people are frustrated, the less they are satisfied. In the beginning of 2014, the cost of

Eurozone solidarity was €135 billion for Germany. So, the Euro skeptic "Alternative für Deutschland" could grow and a German departure from the Euro has a small but sizable probability in five years; a scenario has to be built as "Alternative für Deutschland" which obtained 5% at European and is growing.

2.3.4. Different kinds of systemic banking crisis processes

Scenarios have to be built for many kinds of banking crises; "too big to fail" banks were bailed out by treasuries in the last systemic crisis. AIG, the biggest insurance company, was also bailed out to avoid contagious effects. In 2008–2009, some investment banks and some real-estate credit lending institutions were merged into much bigger financial institutions. These mergers were costly, deceptive and difficult. It is not sure that in the next systemic crisis, there will be financial institutions willing to make that kind of gamble.

The risk aversion of bank treasurers in the 2007 fall led quickly to money market illiquidity. Extraordinary actions by the central bank were necessary to avoid a liquidity crisis. After the Lehman failure, the systemic crisis was stopped by central bank' coordinated actions. This success is very important for the next systemic crisis.

The 1826 systemic banking crisis with runs in London was the first to be solved by a central bank that declared itself a lender of last resort. So, the crisis was over by the next morning. This kind of systemic banking crisis issue has been seen many times since 1826; in July 2012, Draghi pledged that the ECB would do its best to lend in the last resort and this curious sentence chosen to avoid a Bundesbank reaction has been sufficient to induce a rate convergence in the Eurozone. So, this kind of banking systemic crisis can be solved if there is sufficient trust. Even in countries where a protective law on deposits exists, banking failure fears induce runs and panics that imply systemic crisis that the central bank has to fight quickly.

Stress tests can help anticipate solvency problems in banks; recent stress tests showed weaknesses in some medium-sized banks; to avoid failures and contagiousness scenarios, mergers have to be made as soon as possible because it is easier to find buyers before a crisis. The US Treasury had to put a lot of money up to help the buyers of Bear Stern and Merril Lynch in 2008 some months before the Lehman failure.

The probability exists of some dangerous and derived scenarios with different kinds of mean reversing process:

– a systemic banking crisis with contagious fear up to a panic;

– a systemic banking crisis with a market crash;

– a systemic banking crisis with a stock market and/or derivative market crash is the most difficult process to stop because contagion spreads so quickly;

– a systemic banking crisis process with some runs with a systemic financial crisis process because the central bank and/or treasury are not able to stop the gregarious fear process. The media could increase problems;

– a systemic banking crisis with quick panic and stampede too quick to be fought very efficiently could be a disaster with a credit crunch process magnified by the credit multiplier.

It is interesting to focus on a possible systemic scenario which can have an important impact on a five-year projection: the currency union breaks up. Indeed, this potential breakup would be the beginning of a very complex financial crisis and a long restructuring process for the European banking sector as lots of banks would be insolvent. The worst banking crisis ever is a possible scenario, even if there is a strong sense of survival in the banking world and among European governments, but Europe needs stronger governance which can make decisions quickly and efficiently; a better ability to think about future risks could reduce European government reluctance to increase European institutions' power. Many countries are disillusioned by the European Union; northern countries are willing to make investment in the south, not grants and charities; they are disillusioned about southern countries' policy failures. What can be done to solve the various scenarios of banking crisis? In the case of country insolvency linked to banking sector insolvency in a kind of Irish scenario worst case, what can be done? Would partial sovereign debt mutualization be possible? Would special-purpose partial sovereign debt mutualization be possible?

2.4. An efficient five-year scenario generator

A five-year scenario generator with rational expectation hypothesis and efficient market hypothesis can cover classic cycles but not all possibilities.

It is now obvious that the reducing risk trend of the second half of the 20th Century is over. The 21st Century trend is strongly upward. Over five-year scenarios, most operators are acting on implied volatilities and on historic variance, which means that risks trends are not rising; scenario-generator-based measures of risk are much more realistic, taking more precisely risk trends, but it is necessary to define the right scenario generation. A classic generator is based on both rational learning on a classic cycle and more futuristic scenarios resulting from a cycle modified by risk trends, even long risk trends. This does not include the kind of crisis scenarios implying fear-induced irrationalities. Scenarios with endogenous shocks studied in [FRA 12] are included in these rational expectation hypothesis generators.

Working on some 20 basis scenarios to estimate probability-weighted averages and moments makes sense. It is better to multiply this number with stochastic parameter equations. In this kind of generator, 20 basis scenarios are needed over a five-year span of time: the main categories are inflationist scenarios and deflationary scenarios linked or not to some shocks that can occur.

2.4.1. *Combination of two 5-year generators*

For a better parametric stability, two five-year generators are needed.

Timmerman (1993–1996) showed in [TIM 99] that learning implications in standard risk-neutral asset pricing model increases the volatility of asset prices.

Kim showed in [JAN 09] that "adaptive learning can generate the excess volatility, long swings and persistence", which is also quoted in Evans' paper (Frydman and Phelps' book *Rethinking Expectations* [FRY 12]).

Two papers on cycle dynamics (Van Nieuwerburgh Bayesian model learning on productivity [GIA 03]) explain sharp downturns in the classic cycle; generally, learning and crisis remembrances are an important factor for increased volatility and model parameter volatility. Learning means that asset managers will be quicker to recognize patterns and react promptly. So, model parameter volatility can be a problem even in a generator with classic scenarios that are not hit by fear. When a crisis involves fear, irrational behaviors and mistakes, parameter volatility is much more of a problem.

When monetary policy applies the Taylor rule or the "natural rate" Wicksell rule, rates are near the "natural rate"; asset managers and other operators can better understand monetary policy; there is a better stability and model parameters are more stable. Market operators' expectations are less mistaken so that volatility is lower; when the Taylor rule is no longer used, when the central bank manages, on fundamentals, "price targeting", equilibrium is not guaranteed and volatility could be surprising. For model parametric stability of a five-year scenario generator, it is good to use a simple model based on an inflation/deflation rate with few parameters. A five-year scenario generator including many crisis scenarios with some inefficient markets and some irrational expectations could have serious parametric instability problems; therefore, that kind of generator must use few parameters. To have two different generators for an ordinary period and for a crisis process period involves a smaller number of parameters in both generators.

A crisis generator with many scenarios involving fear has to be as simple as possible. When fears spread, many irrational expectations about the kind of crisis can upset markets. It is rational to have rational expectations if most other operators are rational; when expectation of an asset manager is that many operators will be in fear mode and will spread irrationality, it is good to use a specialized crisis scenario generator as soon as possible, when crisis and fear are coming. A rational asset manager's reaction is to reduce risk as quickly as possible which is an old recipe before market illiquidity problems. To pass to crisis mode, by mistake, is costly. This cost can be assessed.

To assume an efficient market hypothesis and rational expectations hypothesis when a fear process is obvious could be more costly. A classic generator that uses the rational expectations hypothesis, the classic economic theory with efficient market hypothesis has few parametric stability problems in quiet periods, but its use during a crisis is dangerous. Medium-term classic scenario generator parameter instability becomes a problem if there are many parameters and a choppy market.

Rational scenarios can be built:

– by extrapolation of short-term trends in one year;

– by extrapolation of the economists' consensus;

– by use of classic cyclical patterns over five years that could be modified by taking into account some obvious trends;

– by near normal cyclical patterns already seen in the past in some countries;

– by use of shocks patterns already seen in the past.

On the contrary, crisis scenarios hit by a fear process have to be based on what experts say: wars, revolutions and sociopolitical events are not possible to forecast but they are possible with a sizable probability, these crisis processes will be addressed by the crisis scenario generator. Irrational scenarios have to be built on a fear behavior process resulting from shocks chosen as hardly to be expected over the five years to come and out of the box! As shocks and abnormal crisis patterns are often very different from previous experience, it could be better to model these scenarios on intuition of fears to divert them from already seen abnormal scenarios, even if, over five years, this kind of abnormal scenario has occurred. Precise value at risk (VaR), conditional value at risk (CVaR) and tail value at risk (TVaR) valuations done with a crisis scenario generator are very useful to manage well during a crisis.

2.4.2. Specific treatment of a social crisis scenario

Camdessus, former IMF head, wrote (2/10/2014 Reform newspaper):

> The characteristic of humanity's history is that crisis is necessary to make reforms possible....During crisis (as 2008) some reforms were made, but after the crisis a return to "Business as usual" is hoped for [...] France is an old and great country more able to make revolution than to adopt reforms.

Revolution means fears and tears scenario processes need to be treated clearly by a crisis generator.

The 1968 student upheavals in France and elsewhere were not fearful scenarios; this surprise social climate change was not possible to forecast and came to a quick end without fear which was not so possible to foresee. This kind of scenario is not to be treated in the crisis generator.

Social crisis could lead to political crisis with fear of a bloody revolution. The French social crisis of 1936 was that kind of fearful scenario which came to an end because President Leon Blum, a clever negotiator, cut a deal which involved the same devaluations of the French Franc as the 1968 deal; nevertheless, scenarios based on the French social crisis of 1936 are to be included in the crisis generator even if this crisis was possible to foresee because it was the result of excessive austerity implying deflation by the Laval government. From this example, it appeared that when people thought that their jobs were no longer at risk because deflation was over, they asked for crisis compensation, ignoring the instability of the economy that they pushed to a new crisis. The same kind of scenario can be found in other Latin countries, but excessive austerity policies driven by the European Union in southern Europe could produce social crisis now. Some deflation scenarios which involve this process are to be treated by a crisis generator because they induce fear.

That kind of crisis can be seen early with some social mood changes that are leading indicators; lead time is short to decide to build scenarios. This kind of crisis, which could be worse than the 1936 crisis which involved increases in cost for industrial product (25–50%) and two devaluations (generally 56%), is to be treated by the crisis generator; some Latin American social crisis scenarios linked to inflation/deflation process were worse and more costly than the 1936 crisis because fear and emotions were prevalent.

Some social upheavals could be seen in Asia, but it is not the kind of fearful scenario that has to be treated in the crisis generator. Slower Chinese growth scenarios (even 5% growth) could be a social problem, but the Chinese development process is not very different from the Japanese development process: 15 years of 10% growth followed by 15 years of 5% growth. The consequences of slower Chinese and European growth on Asian countries could be a bubble bursting and contagious crisis analogous to the 1997–1998 scenarios. Would this rationally imply social crisis? It is possible, but a crisis process that could imply political consequences is not a scenario to be considered; some rational scenarios could be built for classic upheavals without a fear process. All this could mean important consequences on raw materials and raw materials producers in Africa where some bleak scenarios have to be considered. To take the consensus view of economists and even some alternative scenarios by economists, who seem to understand well the kind of unfolding period or analogous period, is rational, although not necessarily sufficient.

In the Middle East, some social events (e.g. food prices, shocks, social upheavals) could not be foreseen; they could be included in a jihad war scenario process; probabilities have to be based on expert opinion. Some heavy shock surprises linked to medium-term war scenarios could be revolutions in Saudi Arabia, Oman and/or the Emirates. Revolutions and subsequent wars in the Middle East would imply major oil price shocks or other "black swans". Worst-case scenarios with very high price for oil and perhaps small probabilities, even over a horizon of five years, could lead to fear contagions that could be as irrational, as dangerous for markets and the world economy.

2.5. Details on several scenarios

Below, we provide some central scenarios for the main countries of each zone.

2.5.1. *Scenarios for the Eurozone*

Now, the most probable one-year scenario for countries that were unable to refinance their debt during the Euro crisis is an extrapolation of rebound. Many kinds of five-year scenarios could result from some unforeseen end of cycle or sociopolitical problems. Better growth and somewhat higher inflation can be seen if some reflation plans with infrastructure works were put in force in Europe.

Nevertheless, over a five-year span of time, with a €300 million Juncker plan [JUN 14], the Euro area growth rate could not be much higher than 1%; over a one-year horizon, 0.6–0.8% could be the limits if the Euro/dollar exchange rate does not move very much. Inflation could be as low as 0.2% with energy prices driven down by the Brent oil price. A rebound to 1% will be seen in less than five years: consequently, a 10 year Bund will be at a rate of 1.2%. Deflation risk can be 20%. Euro frailty to any kind of crisis already quoted can be a problem to many bond markets in Euro except the Bund market. Any crisis will be fought by the ECB which has decided to buy €400 million of private assets by 2016.

Fiscal policy in one year could be characterized by a deficit down to 2.5%; this decrease could go on for the five-year span of time that is foreseeable. Public debt could be more costly to refinance, when US rates' influence will be on the uptrend, but debt to the GNP ratio will be stable at 95%. The current account surplus of the Eurozone could continue to increase as the oil price is down and the Euro is down. The Euro cannot be down very much at a time when the current account is improving. The Euro's lower prices help earnings grow at a 2.5% rate.

Evolution of inflation rates for central scenarios of the five main countries of the Eurozone

Country Nb	County name	Number	2016	2017	2018	2019	2020	End of cycle	Long term tendency	Indicator
2	GERMANY	211	1,2	1,3	1,2	1,2	1,2	Slight economic downturn in 2016	Optimistic	↑
2	GERMANY	212	0,0	0,0	0,2	0,5	0,7	Sharp economic downturn in 2016	Optimistic	↑
2	GERMANY	213	-0,5	-1,0	-0,5	0,0	0,6	Risk of recession in 2016	Optimistic	↑
2	GERMANY	221	0,8	0,6	0,4	0,6	0,9	Slight economic downturn in 2017	Neutral	=
2	GERMANY	222	0,2	0,0	-0,2	0,0	0,5	Sharp economic downturn in 2017	Neutral	=
2	GERMANY	223	-0,5	-1,8	-0,8	0,0	0,4	Risk of recession in 2017	Neutral	=
2	GERMANY	231	0,7	0,6	0,3	0,0	0,2	Slight economic downturn in 2018	Stagnation	→
2	GERMANY	232	0,1	0,0	-0,3	0,2	0,4	Sharp economic downturn in 2018	Stagnation	→
2	GERMANY	233	-0,6	-2,0	-1,0	0,0	1,0	Risk of recession in 2018	Stagnation	→
2	GERMANY	241	0,5	0,4	0,3	0,2	-0,2	Slight economic downturn in 2019	Deflation	↓
2	GERMANY	242	0,4	0,3	0,1	0,0	-1,0	Sharp economic downturn in 2019	Deflation	↓
2	GERMANY	243	-0,7	-2,5	-1,5	-0,3	0,4	Risk of recession in 2019	Deflation	↓
3	FRANCE	111	1,0	1,1	1,0	1,0	1,2	Slight economic downturn in 2016	Optimistic	↑
3	FRANCE	112	0,0	0,0	0,2	0,5	0,7	Sharp economic downturn in 2016	Optimistic	↑
3	FRANCE	113	-0,5	-1,0	-0,5	0,0	0,6	Risk of recession in 2016	Optimistic	↑
3	FRANCE	121	0,8	0,6	0,4	0,6	0,9	Slight economic downturn in 2017	Neutral	=
3	FRANCE	122	0,2	0,0	-0,2	0,0	0,5	Sharp economic downturn in 2017	Neutral	=
3	FRANCE	123	-0,5	-1,8	-0,8	0,0	0,4	Risk of recession in 2017	Neutral	=
3	FRANCE	131	0,7	0,6	0,3	0,0	0,2	Slight economic downturn in 2018	Stagnation	→
3	FRANCE	132	0,1	0,0	-0,3	0,2	0,4	Sharp economic downturn in 2018	Stagnation	→
3	FRANCE	133	-0,6	-2,0	-1,0	0,0	1,0	Risk of recession in 2018	Stagnation	→
3	FRANCE	141	0,5	0,4	0,3	0,2	-0,2	Slight economic downturn in 2019	Deflation	↓
3	FRANCE	142	0,4	0,3	0,1	0,0	-1,0	Sharp economic downturn in 2019	Deflation	↓
3	FRANCE	143	-0,7	-2,5	-1,5	-0,3	0,4	Risk of recession in 2019	Deflation	↓
4	ITALY	111	1,0	1,1	1,0	1,0	1,2	Slight economic downturn in 2016	Optimistic	↑
4	ITALY	112	0,0	0,0	0,2	0,5	0,7	Sharp economic downturn in 2016	Optimistic	↑
4	ITALY	113	-0,5	-1,0	-0,5	0,0	0,6	Risk of recession in 2016	Optimistic	↑
4	ITALY	121	0,8	0,6	0,4	0,6	0,9	Slight economic downturn in 2017	Neutral	=
4	ITALY	122	0,2	0,0	-0,2	0,0	0,5	Sharp economic downturn in 2017	Neutral	=
4	ITALY	123	-0,5	-1,8	-0,8	0,0	0,4	Risk of recession in 2017	Neutral	=
4	ITALY	131	0,7	0,6	0,3	0,0	0,2	Slight economic downturn in 2018	Stagnation	→
4	ITALY	132	0,1	0,0	-0,3	0,2	0,4	Sharp economic downturn in 2018	Stagnation	→
4	ITALY	133	-0,6	-2,0	-1,0	0,0	1,0	Risk of recession in 2018	Stagnation	→
4	ITALY	141	0,5	0,4	0,3	0,2	-0,2	Slight economic downturn in 2019	Deflation	↓
4	ITALY	142	0,4	0,3	0,1	0,0	-1,0	Sharp economic downturn in 2019	Deflation	↓
4	ITALY	143	-0,7	-2,5	-1,5	-0,3	0,4	Risk of recession in 2019	Deflation	↓
5	SPAIN	111	0,8	0,7	0,9	1,5	2,0	Slight economic downturn in 2016	Optimistic	↑
5	SPAIN	112	0,7	0,5	0,6	0,7	0,8	Sharp economic downturn in 2016	Optimistic	↑
5	SPAIN	113	0,6	-0,2	0,0	0,5	0,7	Risk of recession in 2016	Optimistic	↑
5	SPAIN	121	0,5	0,4	0,3	0,7	0,8	Slight economic downturn in 2017	Neutral	=
5	SPAIN	122	0,5	0,4	0,0	0,5	0,7	Sharp economic downturn in 2017	Neutral	=
5	SPAIN	123	0,5	0,0	-0,5	-0,2	0,5	Risk of recession in 2017	Neutral	=
5	SPAIN	131	0,3	0,3	0,2	0,4	0,8	Slight economic downturn in 2018	Stagnation	→
5	SPAIN	132	0,3	0,3	0,0	0,5	0,7	Sharp economic downturn in 2018	Stagnation	→
5	SPAIN	133	0,3	0,3	-0,5	-0,6	0,3	Risk of recession in 2018	Stagnation	→
5	SPAIN	141	0,1	0,2	0,2	0,0	0,2	Slight economic downturn in 2019	Deflation	↓
5	SPAIN	142	0,1	0,1	0,1	-0,4	-0,5	Sharp economic downturn in 2019	Deflation	↓
5	SPAIN	143	0,1	0,1	0,0	-2,0	-2,5	Risk of recession in 2019	Deflation	↓
6	GREECE	111	2,4	2,4	2,5	2,6	2,8	Slight economic downturn in 2016	Optimistic	↑
6	GREECE	112	2,0	2,0	1,8	2,2	2,3	Sharp economic downturn in 2016	Optimistic	↑
6	GREECE	113	-0,3	-0,5	-0,2	0,0	0,3	Risk of recession in 2016	Optimistic	↑
6	GREECE	121	2,3	2,2	2,0	2,2	2,3	Slight economic downturn in 2017	Neutral	=
6	GREECE	122	2,3	2,0	1,5	2,0	2,5	Sharp economic downturn in 2017	Neutral	=
6	GREECE	123	2,3	1,0	-0,5	-0,3	0,0	Risk of recession in 2017	Neutral	=
6	GREECE	131	2,0	2,2	2,1	1,9	2,0	Slight economic downturn in 2018	Stagnation	→
6	GREECE	132	2,0	2,0	1,8	1,0	1,0	Sharp economic downturn in 2018	Stagnation	→
6	GREECE	133	2,0	1,9	-0,8	-1,5	-0,5	Risk of recession in 2018	Stagnation	→
6	GREECE	141	0,9	1,0	0,9	0,7	1,0	Slight economic downturn in 2019	Deflation	↓
6	GREECE	142	0,9	0,8	1,0	0,0	0,5	Sharp economic downturn in 2019	Deflation	↓
6	GREECE	143	0,9	0,8	0,4	-2,0	-1,4	Risk of recession in 2019	Deflation	↓

Evolution of growth rates for central scenarios of the five main countries of the Eurozone

Country Nb	County name	Number	2016	2017	2018	2019	2020	End of cycle	Long term tendency	Indicator
2	GERMANY	211	1,6	1,5	1,8	1,8	2,0	Slight economic downturn in 2016	Optimistic	↑
2	GERMANY	212	1,3	1,0	1,5	1,7	1,9	Sharp economic downturn in 2016	Optimistic	↑
2	GERMANY	213	-0,6	0,5	1,0	2,1	2,0	Risk of recession in 2016	Optimistic	↑
2	GERMANY	221	1,3	1,1	1,3	1,4	1,6	Slight economic downturn in 2017	Neutral	=
2	GERMANY	222	1,3	0,5	0,8	1,0	1,3	Sharp economic downturn in 2017	Neutral	=
2	GERMANY	223	1,3	-1,0	0,7	1,0	1,3	Risk of recession in 2017	Neutral	=
2	GERMANY	231	0,8	1,4	1,0	1,1	1,3	Slight economic downturn in 2018	Stagnation	→
2	GERMANY	232	0,8	1,0	0,3	1,0	1,3	Sharp economic downturn in 2018	Stagnation	→
2	GERMANY	233	0,8	1,2	-1,0	0,8	1,2	Risk of recession in 2018	Stagnation	→
2	GERMANY	241	0,5	0,8	1,4	1,2	1,2	Slight economic downturn in 2019	Deflation	↓
2	GERMANY	242	0,5	0,8	1,0	0,4	0,6	Sharp economic downturn in 2019	Deflation	↓
2	GERMANY	243	0,5	0,7	0,8	-1,0	0,0	Risk of recession in 2019	Deflation	↓
3	FRANCE	111	1,1	0,8	1,1	1,3	1,3	Slight economic downturn in 2016	Optimistic	↑
3	FRANCE	112	0,5	0,7	0,9	1,1	1,3	Sharp economic downturn in 2016	Optimistic	↑
3	FRANCE	113	-0,5	0,4	1,0	1,2	1,3	Risk of recession in 2016	Optimistic	↑
3	FRANCE	121	0,9	1,2	0,8	1,2	1,2	Slight economic downturn in 2017	Neutral	=
3	FRANCE	122	0,9	0,4	0,8	1,2	1,2	Sharp economic downturn in 2017	Neutral	=
3	FRANCE	123	0,9	-0,7	0,5	1,2	1,2	Risk of recession in 2017	Neutral	=
3	FRANCE	131	0,8	1,0	1,2	0,8	0,9	Slight economic downturn in 2018	Stagnation	→
3	FRANCE	132	0,8	0,9	0,2	0,8	0,9	Sharp economic downturn in 2018	Stagnation	→
3	FRANCE	133	0,8	0,9	-0,8	0,8	0,9	Risk of recession in 2018	Stagnation	→
3	FRANCE	141	0,6	1,0	1,4	1,5	0,5	Slight economic downturn in 2019	Deflation	↓
3	FRANCE	142	0,6	0,7	0,8	0,0	0,5	Sharp economic downturn in 2019	Deflation	↓
3	FRANCE	143	0,6	0,6	0,6	-1,0	0,5	Risk of recession in 2019	Deflation	↓
4	ITALY	111	0,9	0,6	0,7	1,0	1,2	Slight economic downturn in 2016	Optimistic	↑
4	ITALY	112	0,5	0,6	0,7	0,9	1,2	Sharp economic downturn in 2016	Optimistic	↑
4	ITALY	113	-0,8	0,4	0,8	1,0	1,2	Risk of recession in 2016	Optimistic	↑
4	ITALY	121	1,2	0,7	0,6	0,8	1,0	Slight economic downturn in 2017	Neutral	=
4	ITALY	122	1,2	0,0	0,2	0,5	1,0	Sharp economic downturn in 2017	Neutral	=
4	ITALY	123	1,2	-1,0	0,0	0,5	1,0	Risk of recession in 2017	Neutral	=
4	ITALY	131	0,7	0,9	0,6	0,8	0,9	Slight economic downturn in 2018	Stagnation	→
4	ITALY	132	0,7	0,7	0,3	0,6	0,9	Sharp economic downturn in 2018	Stagnation	→
4	ITALY	133	0,7	0,8	-1,0	0,4	0,9	Risk of recession in 2018	Stagnation	→
4	ITALY	141	0,6	0,8	0,8	0,6	0,7	Slight economic downturn in 2019	Deflation	↓
4	ITALY	142	0,6	0,7	0,7	0,0	0,5	Sharp economic downturn in 2019	Deflation	↓
4	ITALY	143	0,6	0,6	0,6	-1,1	0,3	Risk of recession in 2019	Deflation	↓
5	SPAIN	111	1,0	1,2	1,4	1,6	1,7	Slight economic downturn in 2016	Optimistic	↑
5	SPAIN	112	0,7	1,0	1,3	1,6	1,7	Sharp economic downturn in 2016	Optimistic	↑
5	SPAIN	113	-0,5	0,5	1,0	1,5	1,7	Risk of recession in 2016	Optimistic	↑
5	SPAIN	121	1,2	1,0	1,2	1,4	1,4	Slight economic downturn in 2017	Neutral	=
5	SPAIN	122	1,2	0,6	1,0	1,2	1,4	Sharp economic downturn in 2017	Neutral	=
5	SPAIN	123	1,2	-1,0	0,5	1,0	1,4	Risk of recession in 2017	Neutral	=
5	SPAIN	131	1,1	0,9	1,1	1,2	1,2	Slight economic downturn in 2018	Stagnation	→
5	SPAIN	132	1,1	0,6	0,9	1,0	1,2	Sharp economic downturn in 2018	Stagnation	→
5	SPAIN	133	1,1	1,1	-1,0	0,5	1,0	Risk of recession in 2018	Stagnation	→
5	SPAIN	141	1,0	0,7	0,9	0,9	1,0	Slight economic downturn in 2019	Deflation	↓
5	SPAIN	142	1,0	0,3	0,6	0,8	1,0	Sharp economic downturn in 2019	Deflation	↓
5	SPAIN	143	1,0	0,9	0,9	-1,0	0,5	Risk of recession in 2019	Deflation	↓
6	GREECE	111	2,0	2,2	2,3	2,3	2,4	Slight economic downturn in 2016	Optimistic	↑
6	GREECE	112	0,6	0,8	1,0	1,2	1,4	Sharp economic downturn in 2016	Optimistic	↑
6	GREECE	113	-2,0	-1,0	0,0	0,5	1,2	Risk of recession in 2016	Optimistic	↑
6	GREECE	121	2,4	2,0	1,9	2,0	2,0	Slight economic downturn in 2017	Neutral	=
6	GREECE	122	2,4	1,0	1,2	1,4	1,4	Sharp economic downturn in 2017	Neutral	=
6	GREECE	123	2,4	-2,5	-0,5	0,5	1,0	Risk of recession in 2017	Neutral	=
6	GREECE	131	1,2	1,0	0,8	1,2	1,2	Slight economic downturn in 2018	Stagnation	→
6	GREECE	132	1,2	0,8	0,3	0,7	1,0	Sharp economic downturn in 2018	Stagnation	→
6	GREECE	133	1,2	1,0	-3,0	-1,5	0,0	Risk of recession in 2018	Stagnation	→
6	GREECE	141	0,2	0,3	0,0	0,3	0,3	Slight economic downturn in 2019	Deflation	↓
6	GREECE	142	0,2	0,0	0,2	0,0	0,2	Sharp economic downturn in 2019	Deflation	↓
6	GREECE	143	0,2	0,0	-0,9	-3,0	-2,0	Risk of recession in 2019	Deflation	↓

France: most probable one-year scenarios

Very slow growth with momentary stagnation is the most probable scenario; very subdued inflation could mean deflationary risk. What surprise could be seen beside a negative inflation rate? Which other one-year scenarios could be seen? Oil price drop and commodities' rather weak prices, what kind of growth surprise could be faced? A scenario with a significant growth surprise is possible.

France: most probable five-year scenarios

The fundamental problem of France is the public sector cost: 57% GNP compared to 27% in 1973; at this time, debt to GNP was 20% compared to a ratio near 100% today. Can this trend be reversed in a five-year scenario? There is a low probability of this trend being reversed.

Five-year scenarios are numerous. As in Spain and Italy, political scenarios are very meaningful. The presidential election in May 2017 could be a big change. Many reforms made from 2007 to 2012 were erased in the following years. Reforms to be made on the economy: to boost competition in oligopolistic sectors, energy, for example, in retail and other networks and reforms to labor market as decentralization bargaining process in branch collective agreements are not easy to implement. The Macron act [MAC 14] was reduced by opponents' actions. So, a very slow process is the most probable reform scenario. In May 2017, the presidential election will be followed by a general election. Usually, when two elections are held in a short space of time, the results are not inconsistent. So, a victory for the left could mean more of the same; a victory for the center right could mean, lower taxes, more.

France: other possible scenarios

Some other scenarios are possible for France, particularly scenarios with still increasing budget deficits linked to military operations and consequently police needs. Even if Brussels and Berlin are putting on pressures to limit the French debt to GNP ratio, these scenarios have to be considered with interest. Decreasing the budget deficit somewhat is also possible.

Germany: most probable one-year scenario

On the one year basis, approximately 2% growth with 1% inflation linked to the Euro decreasing exchange rate and better credit availability is an optimist scenario with a bigger probability than other such scenarios.

Less optimistic scenarios need to be considered with exports to Russia being weak as long as energy prices are low, and as long as conflicts with European countries imply punitive sanctions.

Nevertheless, the German people's confidence is much higher than that of people of other Eurozone countries where unemployment is higher; German domestic demand is stronger than usual because minimum wage negotiations raised income for many low-income Germans, but weaker demand for German products from abroad could reduce the current account surplus, confidence level and domestic consumption growth. If scenarios with weaker growth and weaker inflation are more probable, scenarios have to be built as a function of the ECB's quantitative easing amount; the infrastructure plan amount is also an important parameter to differentiate five-year scenarios. In a country where debt to GNP ratio is beginning to decrease with a budget being balanced, even if a big infrastructure investment program is decided, confidence is helped.

Germany: most probable five-year scenarios

On the five years basis, optimistic scenarios are really linked to ECB quantitative easing amounts, especially organized to boost credit. A rebound is possible because wages are rising, not only linked to minimum wage deals, but also to economic activity; construction activity is already helped by low interest rates; when the infrastructure investment program adds activity, a decrease in unemployment (5.5% in 2012, 5% in 2014 and perhaps 4.7% in 2015) could speed up; manpower shortage could mean, in this aging country, some wage increases. Wage increases are necessary to lure foreign engineers and other graduate migrants by attractive opportunities in German industry several percent in 2015).

Figure 2.2. *IFO barometer for Germany in trade and industry*

Over this five-year horizon, a start of investment cycle seems probable because some well-known economists alleged that export market shares could be lost if research and investment budgets are not increased; an industry refurbishing plan could be made, but it could be a lengthy process. A better demand for exports could help. Scenarios can be built on the assumption of a growth in demand from Asia, eastern Europe and southern European states; from southern Europe, exports are growing (5%) and will accelerate in optimistic scenarios; therefore, current account surplus would remain at a high level (7–7.5%) and will limit the Euro exchange rate decline. German growth forecasts have to be based on domestic consumption and export demand. German exports are influenced by Chinese demand which will probably grow at 7% for some years, other Asian export demands which could grow significantly in the coming years, but also by European demand. As a consequence, German export forecasts, which have an important impact on growth, are really difficult and require many scenarios.

Germany: other less optimistic scenarios

In case of European stagnation and deflation, lower growth is also possible (some recent figures were surprisingly weak as the second quarter growth figure was negative) and leading indicators were flat; in this case, inflation would be lower than 1% in spite of some wage increases linked to demographic problems. Stagnation would be a worst case with zero inflation with a sizable probability. This probability could be increased by the Ukrainian problems becoming worse and spreading; Russian demand for German exports could diminish; but Russian imports from Germany only account for approximately 3% of total German exports; it is a pity that an unstable geopolitical situation is a hindrance for an investment cycle to start.

Italy: most probable scenario

The political situation gave to Matteo Renzi a leadership to make some structural reforms (judicial power, Senate reform and worker administration regulations in particular) but they are not easy to implement; Matteo Renzi's leadership is subsiding after one year of government with disappointing growth and big difficulties in implementing labor market reform which have produced sociopolitical demonstrations. If there are negative figures for growth by the end of 2014 and later, his popularity and leadership will be much reduced. In spite of negative figures for growth, budgetary deficit is stabilized lower than 3% and debt to GNP ratio is nearly stabilized at 130%. Consumption is stable and credit availability is beginning to improve;

exports are still growing quicker than imports. Growth is badly needed for Italy to avoid deflation and to be on the way out of crisis; a small inflation rebound from 0.3 to 0.5% reduces deflation risks.

Over a one-year horizon, it is possible if Matteo Renzi is able to win in the elections that could be held in summer 2015 that he could introduce adequate reforms. This is the most probable scenario, but some other scenarios are also possible.

Italy: other one-year scenarios

Stagnation and deflationary scenarios are possible. These economic scenarios lead to more deficit and more debt and less leadership for Matteo Renzi, even if the summer 2015 elections are helpful for him; another kind of scenario would appear if Renzi has to resign: political instability with very small figures for growth and inflation.

An optimistic five-year scenario could be possible if Matteo Renzi is able to achieve his reform program; Italian economy convalescence could be seen. However, pessimistic scenarios based on incomplete reforms and political instability are the most probable. Note that small hazardous figures for growth and inflation could imply difficult debt refinance.

Spain: most probable scenarios

Production rebound watched for a year is quicker during the last few months on a trend of 3% with a stronger consumption and employment. Budget deficit is down from 3.6 to 3.1%. With this kind of better economy, the next election that will take place in 2015 could be possible to win for Mariano Rajoy in spite of criticism and the Catalonia referendum risk. This province is so much in debt that its debt would be very difficult to refinance if Catalonia were an independent state. As with Scotland, Catalonia could obtain financial advantages due to a referendum threat, even if this referendum is not compatible with the constitution. Spanish exports have been, up to now, a success showing a good competitiveness; it is possible that trade balance rebound could be over because imports driven by consumption have a rebound and goods available to export are less numerous when the Spanish consumer is more active.

These impressive results could go on or stall because the regional election in May 2015 and general election later this year could mean an unstable government as the third party "Podemos" is nearly as important as the two main parties, having surged from "indignados", a far left movement; therefore, there are three scenarios: even, after five years of recession and some scandals, Rajoy could be in power because unemployment has begun to decrease from 24.7 to 23.7% in one year. With an accelerating growth from 1.2% in 2014 to a forecast of 2% for 2015, in spite of a five-year recession and a recent government expenditure decrease, Rajoy could be re-elected, but it is an uphill battle. If this battle is won, a five-year positive scenario could be most probable; so Spain, as Ireland, could show that there is a possible convalescence for countries willing to follow IMF advice.

Spain: other one-year scenarios

Other one-year scenarios are possible to devise: for example, a coalition of the Spanish socialist party (PSOE) with Podemos could replace the Mariano Rajoy government in the next general election. The worst-case scenario could result from a very weak PSOE because the "Podemos" program is leftist, as with the Greek government program, with nationalizations, earlier retirement, minimum wage increase and so on; these people are unable to understand the real world, the global world.

Spain: probable five-year scenarios

Five-year pessimistic scenarios can be devised on the upsurge of Podemos. PSOE with Zapatero followed IMF advice, so very many unemployed, especially young, people scorn PSOE, thinking "Podemos = we can do it". They can provoke a disaster and a sociopolitical crisis that could turn into the revolution of "Indignados". Since 2010, "Indignados" have led many important political demonstrations with many hundreds of thousands of people on Paseo del Prado in Madrid and in many towns, with a 1968 tone, with many leftist catchphrases on banners.

Social crisis can be contagious in Latin countries, especially when economy is on the way to convalescence after a long crisis.

A better scenario would be a PSOE government that could mean a classic reflation; if this new government could try reflation the kind of reflation that the IMF advises, it would not be so bad.

Portugal: most probable scenario

This country complied very precisely with advised reforms which were well achieved in spite of opposition; results are coming in with a growth rebound for the last 18 months; forecasted growth for 2015 is 1.9% so that debt refinance is no longer a problem; moreover, Portugal did not ask for recurring help from the IMF and the European Union. In spite of some austerity measures rejected by the constitutional court, budget deficit is stable, unemployment is decreasing and indicators are positive.

The most probable one-year scenario would be that this country will go on applying IMF advice with a moderate relation: income revenue rate is lowered from 23 to 21% and minimum wage will be moderately increased; therefore, growth rate would increase from 1% in 2014 to 1.5% in 2015; unemployment is down from 14.8 to 13.4%; it could go on for a year or five years.

Portugal: other scenarios

This convalescent country could be upset by a social crisis in Spain that would be the worst-case scenario over the one- and five-year horizon. Another bad scenario would be deflation contagion from Europe as zero inflation could mean contagious frailty.

Optimistic scenarios are possible with a growth of 2% and an inflation of 0.6% or higher in 2015.

Ireland: most probable scenario

Growth going on as recently (catch up at 6.5% per year would be followed by 3.8 to 4% in 2015 and 3.3 to 3.6% in 2016); these figures, on average, of a one-year scenario and a five-year scenario imply the highest levels since 2007 with investment cycle on its rebound of 9% and exports helping growth on a 7% trend; therefore, Standard & Poor's and Moody's ratings can be increased, which is a help to refinance debt with lower rates. With these figures (stable inflation at a rate of 0.6%) and continuing to stabilize the banking sector, Ireland will be on the way out of difficulties.

Ireland: other scenarios

A five-year scenario with stronger figures with 1% inflation could lead to excesses but beyond the one-year horizon; another scenario with excesses during this one-year period is not impossible, but not very probable. Scenarios with smaller growth figures are very possible because a rebound could subside in a gradually tempering process; this rebound could end in the case of a worldwide end of cycle or new Euro crisis.

Greece: most probable scenario

Some growth is back on the second quarter of 2014, which is a first result after a five-year recession; leading indicators are on an upward track.

However, the new government has been elected on a program which is a pipe dream. As a consequence, the increase of growth observed during the second quarter of 2014 could cease.

Greece: other scenarios

In southern countries, it is a classic pattern: when the economy is emerging from a deep and protracted recession, social and political turmoil represents a really difficult problem with many possible scenarios. Some growth came back in the second quarter of 2014; it is the first time for the last five years. Leading indicators are up and treasury revenues are up too; tourism is increasing as Islamic counties are becoming riskier; deficits are lower than forecasted figures; but the help of the IMF and the European Union will still be badly needed to refinance foreign debt; debt to GNP ratio is 180%, but the average life of Greek bonds is rather long term; as the very leftist government promised, to be elected, the Moon, the Earth, that is to exact the end of austerity measures, a new crisis could be seen; negotiations between Angela Merkel and the head of Greek government Alexis Tsipras are very difficult. The IMF is willing to lend some small amounts in exchange for reform pledges to avoid for some weeks or months a Greek government failure; this time of negotiation could not last very long because some anger is already seen in Germany and in Greece where calls for more help is thought of as normal. An upheaval worst case scenario could be seen; many kinds of "Grexit": Greece out of the Eurozone could be seen, but the worst is never sure. Some scenarios of a new partial debt annulment could be seen.

2.5.2. Scenarios for English-speaking countries

Evolution of inflation rates for the United Kingdom and the USA

Country Nb	County name	Number	2016	2017	2018	2019	2020	End of cycle	Long term tendency	Indicator
7	UK	111	2,3	2,1	2,3	2,5	2,7	Slight economic downturn in 2016	Optimistic	↑
7	UK	112	1,8	1,5	1,8	2,3	2,5	Sharp economic downturn in 2016	Optimistic	↑
7	UK	113	1,2	1,0	1,5	1,7	2,0	Risk of recession in 2016	Optimistic	↑
7	UK	121	2,5	2,4	2,1	2,3	2,5	Slight economic downturn in 2017	Neutral	=
7	UK	122	2,5	2,4	1,5	1,8	2,0	Sharp economic downturn in 2017	Neutral	=
7	UK	123	2,5	1,0	1,4	1,6	1,7	Risk of recession in 2017	Neutral	=
7	UK	131	2,2	2,1	2,0	2,2	2,0	Slight economic downturn in 2018	Stagnation	→
7	UK	132	2,2	2,0	1,6	1,3	1,6	Sharp economic downturn in 2018	Stagnation	→
7	UK	133	2,2	1,2	0,0	0,2	0,3	Risk of recession in 2018	Stagnation	→
7	UK	141	2,0	2,2	2,0	1,8	2,2	Slight economic downturn in 2019	Deflation	↓
7	UK	142	2,0	2,0	1,8	1,0	1,2	Sharp economic downturn in 2019	Deflation	↓
7	UK	143	2,0	1,8	0,9	-0,4	0,0	Risk of recession in 2019	Deflation	↓
8	USA	111	2,3	2,3	2,5	2,6	2,7	Slight economic downturn in 2016	Optimistic	↑
8	USA	112	1,9	1,4	1,6	2,0	2,5	Sharp economic downturn in 2016	Optimistic	↑
8	USA	113	1,8	1,0	1,4	1,9	2,4	Risk of recession in 2016	Optimistic	↑
8	USA	121	2,6	2,4	2,5	2,6	2,6	Slight economic downturn in 2017	Neutral	=
8	USA	122	2,6	2,0	2,3	2,4	2,5	Sharp economic downturn in 2017	Neutral	=
8	USA	123	2,6	2,4	1,0	1,2	1,9	Risk of recession in 2017	Neutral	=
8	USA	131	2,0	2,0	2,0	2,0	2,5	Slight economic downturn in 2018	Stagnation	→
8	USA	132	2,0	1,9	1,6	1,8	2,3	Sharp economic downturn in 2018	Stagnation	→
8	USA	133	2,0	2,2	2,0	1,3	1,0	Risk of recession in 2018	Stagnation	→
8	USA	141	1,7	1,7	1,8	1,6	1,7	Slight economic downturn in 2019	Deflation	↓
8	USA	142	1,7	1,8	1,8	1,4	1,5	Sharp economic downturn in 2019	Deflation	↓
8	USA	143	1,7	1,9	2,1	1,3	1,0	Risk of recession in 2019	Deflation	↓

Evolution of growth rates for the United Kingdom and the USA

Country Nb	County name	Number	2016	2017	2018	2019	2020	End of cycle	Long term tendency	Indicator
7	UK	111	2,0	2,2	2,0	1,8	2,0	Slight economic downturn in 2016	Optimistic	↑
7	UK	112	1,6	2,0	2,2	2,0	2,0	Sharp economic downturn in 2016	Optimistic	↑
7	UK	113	0,3	1,2	1,9	2,2	2,0	Risk of recession in 2016	Optimistic	↑
7	UK	121	2,2	1,9	2,2	2,2	2,0	Slight economic downturn in 2017	Neutral	=
7	UK	122	2,2	1,5	2,0	2,2	2,0	Sharp economic downturn in 2017	Neutral	=
7	UK	123	2,2	-0,2	1,5	1,8	2,0	Risk of recession in 2017	Neutral	=
7	UK	131	2,0	1,8	2,0	2,2	1,8	Slight economic downturn in 2018	Stagnation	→
7	UK	132	2,0	1,4	1,8	1,9	1,8	Sharp economic downturn in 2018	Stagnation	→
7	UK	133	2,0	2,4	-0,3	1,0	1,6	Risk of recession in 2018	Stagnation	→
7	UK	141	1,8	1,6	1,8	1,5	1,6	Slight economic downturn in 2019	Deflation	↓
7	UK	142	1,8	0,9	1,2	1,2	1,6	Sharp economic downturn in 2019	Deflation	↓
7	UK	143	1,8	2,0	1,8	-0,5	1,0	Risk of recession in 2019	Deflation	↓
8	USA	111	3,0	3,3	3,2	3,3	3,2	Slight economic downturn in 2016	Optimistic	↑
8	USA	112	1,5	1,9	2,5	3,1	3,0	Sharp economic downturn in 2016	Optimistic	↑
8	USA	113	-0,2	3,0	3,5	3,3	3,5	Risk of recession in 2016	Optimistic	↑
8	USA	121	3,3	3,0	3,3	3,5	3,0	Slight economic downturn in 2017	Neutral	=
8	USA	122	3,3	2,3	3,0	3,2	3,0	Sharp economic downturn in 2017	Neutral	=
8	USA	123	3,3	-0,5	1,5	2,8	3,0	Risk of recession in 2017	Neutral	=
8	USA	131	2,0	1,8	2,0	2,5	2,9	Slight economic downturn in 2018	Stagnation	→
8	USA	132	2,0	2,3	1,0	2,0	2,8	Sharp economic downturn in 2018	Stagnation	→
8	USA	133	2,0	2,5	-1,0	1,5	2,5	Risk of recession in 2018	Stagnation	→
8	USA	141	1,0	0,8	1,0	1,3	2,0	Slight economic downturn in 2019	Deflation	↓
8	USA	142	1,0	0,5	0,8	1,0	0,8	Sharp economic downturn in 2019	Deflation	↓
8	USA	143	1,0	1,2	1,5	-2,0	-0,5	Risk of recession in 2019	Deflation	↓

The United Kingdom: most probable scenario

As in some other European countries, the general election to be held in 2015 will change the set of five-year scenarios. The Labour Party seem likely to win, but now the UK Independence Party (UKIP) could be higher than 15%, the Liberal Party could be approximately 10% and the Scottish National Party and the Ulster Unionist Party could take many votes from Labour and from the Conservatives; therefore, on 7 May, the two-party rule could be over; there could be numerous governmental instability scenarios with various consequences as in the period 1974–1979.

The one-year optimistic scenario, nevertheless, is not bad: a sturdy growth (3%) is increasing labor demand but did not start a wage rise; employment ratio is already at the highest level since 1975, but low-paid jobs are an increasing proportion of total employment; unemployment was at a high level in 2011, it is 6.5% in 2014 and could be near 5%, one year from now. So, rates could continue to rise (1.5% at the horizon of one year).

A pessimistic scenario could be linked to political uncertainty with lower growth (2.5% in 2015), very high proportion of low stability new jobs; unemployment being, nevertheless, stable at 6.5%. So, rates would remain below 1.5%.

The United Kingdom: five-year scenarios

The five-year scenario with governmental instability and a vocal UKIP could be the way out of the European Union, but it is a worst-case scenario; worst case is never sure and could have a low probability.

A surprise scenario, with Cameron regaining the British people's confidence could mean a higher growth rate, wage rise, lower savings rate and quicker interest rate rise. This low probability scenario could also be a way out of the European Union.

There is a more probable scenario with a coalition government working to avoid the way out of the European Union. The political scene could be a problem concerning the plan size of economic reflation as a lower growth rate was seen at the beginning of 2015, especially in Scotland.

Low inflation would no longer be a problem as trade unions would push wages slightly higher before a weak government.

USA: most probable one-year scenario

Composite leading indicators are pointing to a continued growth for the USA and Canada; the most probable scenario is a moderate growth (2.5%) with long-term rates at 2.5%. Quoted company earnings could be increasing, still on a 8% trend, so investment cycle could go on as enterprises are cash rich with old equipment (private sector fixed capital has an average age of 22 years). A Republican majority at Congress could reduce expense resulting from Obama's laws on health and on climate in the next budget being discussed in April/May 2015. Budget problems between Republicans and Democrats could be difficult at a time when the debt ceiling has to be set again; some disruptions are possible, but 18 months before a presidential election, congressmen are cautious about opinion reactions to controversies.

USA: other one-year scenario

A more optimistic scenario based on a quicker beginning of the equipment cycle and better exports result for US industry could imply a 3.5% growth, some wage increases, lower unemployment and employment benefits. As the stock market is rather expensive, acquisitions will be fewer and CAPEX will develop. The real estate market with low mortgage rates and rents going up is set for acceleration; demographics are very favorable. Solid growth in the best stage of the cycle for companies' earnings is the happy time before wage increases and big CAPEX expenditures.

USA: most probable five-year scenario

Rather strong growth could be seen in the USA (3%) with an investment cycle in expansion and stronger earnings trend, but rates would begin a cyclical up move. The dollar exchange rates would go on an uptrend; this kind of situation could imply a burst of enthusiasm and the end of cycle, say, before the end of the five-year scenario's span of time.

A better scenario is not usual with a longer cycle; thus, it is a lower probability scenario with a slower growth as seen since 2009, a slower house building scenario, a slower CAPEX cycle; therefore, the usual American burst of enthusiasm would precipitate the end of cycle that could last longer than the five-year scenarios, into the "2020s". Risk aversion would not come back quickly as the great recession will be more than 10 years before the next recession at a time when US public debt could be lower. Public debt could remain just below the 2016 GNP figure as a ceiling and edge away from this figure in the early 2020s.

The reverse scenario of an early end of cycle or a weak growth or stagnation could be seen if an external event (e.g. never seen climatic) hindering growth. That kind of cycle end of scenario does not have a sizable probability.

2.5.3. Scenarios for Asia

Evolution of inflation rates for the main Asian countries

Country Nb	County name	Number	2016	2017	2018	2019	2020	End of cycle	Long term tendency	Indicator
9	CHINA	111	2,5	2,3	2,5	2,7	3,0	Slight economic downturn in 2016	Optimistic	↑
9	CHINA	112	2,0	1,0	1,5	2,0	3,0	Sharp economic downturn in 2016	Optimistic	↑
9	CHINA	113	1,0	0,0	1,0	2,0	3,5	Risk of recession in 2016	Optimistic	↑
9	CHINA	121	2,8	2,7	2,5	2,8	3,0	Slight economic downturn in 2017	Neutral	=
9	CHINA	122	2,8	2,8	1,6	2,0	2,5	Sharp economic downturn in 2017	Neutral	=
9	CHINA	123	2,8	3,0	1,3	2,0	2,7	Risk of recession in 2017	Neutral	=
9	CHINA	131	2,2	2,1	1,9	2,0	2,3	Slight economic downturn in 2018	Stagnation	→
9	CHINA	132	2,2	2,1	1,4	1,8	2,2	Sharp economic downturn in 2018	Stagnation	→
9	CHINA	133	2,2	2,0	1,0	1,5	2,3	Risk of recession in 2018	Stagnation	→
9	CHINA	141	1,0	0,9	0,7	0,8	1,0	Slight economic downturn in 2019	Deflation	↓
9	CHINA	142	1,0	1,2	1,0	0,4	0,8	Sharp economic downturn in 2019	Deflation	↓
9	CHINA	143	1,0	1,5	2,0	0,0	1,0	Risk of recession in 2019	Deflation	↓
10	JAPAN	111	1,3	1,2	1,8	2,0	2,5	Slight economic downturn in 2016	Optimistic	↑
10	JAPAN	112	1,0	0,7	1,0	2,2	2,7	Sharp economic downturn in 2016	Optimistic	↑
10	JAPAN	113	0,5	1,0	0,6	1,0	2,0	Risk of recession in 2016	Optimistic	↑
10	JAPAN	121	1,9	1,9	1,7	2,4	2,6	Slight economic downturn in 2017	Neutral	=
10	JAPAN	122	1,9	1,8	1,5	1,8	2,4	Sharp economic downturn in 2017	Neutral	=
10	JAPAN	123	1,9	1,6	0,0	-0,5	1,0	Risk of recession in 2017	Neutral	=
10	JAPAN	131	1,6	1,6	1,4	1,3	1,6	Slight economic downturn in 2018	Stagnation	→
10	JAPAN	132	1,6	1,5	1,1	0,5	1,0	Sharp economic downturn in 2018	Stagnation	→
10	JAPAN	133	1,6	1,0	0,0	-1,0	0,0	Risk of recession in 2018	Stagnation	→
10	JAPAN	141	1,2	1,0	1,2	1,4	1,8	Slight economic downturn in 2019	Deflation	↓
10	JAPAN	142	1,0	0,5	0,8	1,0	1,5	Sharp economic downturn in 2019	Deflation	↓
10	JAPAN	143	0,2	0,0	-0,2	-1,5	-0,5	Risk of recession in 2019	Deflation	↓
13	INDIA	111	5,5	5,4	6,0	7,0	8,0	Slight economic downturn in 2016	Optimistic	↑
13	INDIA	112	5,0	4,5	5,5	6,6	7,8	Sharp economic downturn in 2016	Optimistic	↑
13	INDIA	113	2,0	1,5	3,0	6,0	7,0	Risk of recession in 2016	Optimistic	↑
13	INDIA	121	6,4	6,2	6,0	7,0	7,5	Slight economic downturn in 2017	Neutral	=
13	INDIA	122	6,4	5,4	5,0	5,5	6,5	Sharp economic downturn in 2017	Neutral	=
13	INDIA	123	6,4	5,0	3,0	4,0	5,0	Risk of recession in 2017	Neutral	=
13	INDIA	131	6,0	5,7	6,0	7,0	6,0	Slight economic downturn in 2018	Stagnation	→
13	INDIA	132	6,0	4,0	2,0	3,0	5,0	Sharp economic downturn in 2018	Stagnation	→
13	INDIA	133	6,0	3,0	0,0	-0,2	1,0	Risk of recession in 2018	Stagnation	→
13	INDIA	141	4,0	5,0	4,0	3,0	4,0	Slight economic downturn in 2019	Deflation	↓
13	INDIA	142	4,0	3,0	2,5	2,0	3,0	Sharp economic downturn in 2019	Deflation	↓
13	INDIA	143	4,0	3,0	2,0	-0,5	-1,5	Risk of recession in 2019	Deflation	↓

Evolution of growth rates for the main Asian countries

Country Nb	County name	Number	2016	2017	2018	2019	2020	End of cycle	Long term tendency	Indicator
9	CHINA	111	6,3	6,8	6,5	6,0	5,8	Slight economic downturn in 2016	Optimistic	↑
9	CHINA	112	5,5	6,0	6,3	6,0	5,8	Sharp economic downturn in 2016	Optimistic	↑
9	CHINA	113	2,0	4,0	5,0	5,5	5,8	Risk of recession in 2016	Optimistic	↑
9	CHINA	121	7,0	6,5	6,8	6,0	5,5	Slight economic downturn in 2017	Neutral	=
9	CHINA	122	7,0	5,5	6,2	6,0	5,5	Sharp economic downturn in 2017	Neutral	=
9	CHINA	123	7,0	1,0	3,0	5,0	5,5	Risk of recession in 2017	Neutral	=
9	CHINA	131	6,7	6,5	6,0	5,0	5,0	Slight economic downturn in 2018	Stagnation	→
9	CHINA	132	6,5	5,5	5,8	5,4	5,0	Sharp economic downturn in 2018	Stagnation	→
9	CHINA	133	6,3	5,0	0,0	4,0	4,0	Risk of recession in 2018	Stagnation	→
9	CHINA	141	5,8	5,3	5,5	4,8	4,0	Slight economic downturn in 2019	Deflation	↓
9	CHINA	142	5,5	4,7	5,0	4,0	3,0	Sharp economic downturn in 2019	Deflation	↓
9	CHINA	143	4,0	3,5	3,0	0,0	0,2	Risk of recession in 2019	Deflation	↓
10	JAPAN	111	2,5	2,9	2,5	2,2	2,0	Slight economic downturn in 2016	Optimistic	↑
10	JAPAN	112	2,0	2,5	2,5	2,2	2,0	Sharp economic downturn in 2016	Optimistic	↑
10	JAPAN	113	-1,0	0,5	1,0	1,5	2,0	Risk of recession in 2016	Optimistic	↑
10	JAPAN	121	1,9	2,0	1,8	1,6	1,5	Slight economic downturn in 2017	Neutral	=
10	JAPAN	122	1,9	1,5	1,8	1,6	1,5	Sharp economic downturn in 2017	Neutral	=
10	JAPAN	123	1,9	-1,5	0,0	0,7	1,5	Risk of recession in 2017	Neutral	=
10	JAPAN	131	1,5	1,6	1,4	1,2	1,0	Slight economic downturn in 2018	Stagnation	→
10	JAPAN	132	1,5	1,5	0,8	1,0	1,0	Sharp economic downturn in 2018	Stagnation	→
10	JAPAN	133	1,5	1,5	-2,0	0,0	1,0	Risk of recession in 2018	Stagnation	→
10	JAPAN	141	1,3	1,1	1,2	1,0	0,8	Slight economic downturn in 2019	Deflation	↓
10	JAPAN	142	1,3	1,3	1,2	0,6	0,6	Sharp economic downturn in 2019	Deflation	↓
10	JAPAN	143	1,3	1,2	1,0	-2,5	0,2	Risk of recession in 2019	Deflation	↓
13	INDIA	111	6,2	6,8	7,5	8,0	8,5	Slight economic downturn in 2016	Optimistic	↑
13	INDIA	112	5,1	6,0	7,0	8,0	8,5	Sharp economic downturn in 2016	Optimistic	↑
13	INDIA	113	3,0	5,0	6,0	7,6	8,3	Risk of recession in 2016	Optimistic	↑
13	INDIA	121	6,6	6,2	6,7	7,8	8,2	Slight economic downturn in 2017	Neutral	=
13	INDIA	122	6,6	4,0	5,5	7,5	8,0	Sharp economic downturn in 2017	Neutral	=
13	INDIA	123	6,6	1,0	3,0	5,0	7,0	Risk of recession in 2017	Neutral	=
13	INDIA	131	6,0	7,0	6,0	6,4	6,8	Slight economic downturn in 2018	Stagnation	→
13	INDIA	132	6,0	5,0	4,0	5,0	6,0	Sharp economic downturn in 2018	Stagnation	→
13	INDIA	133	6,0	7,0	-0,2	3,0	5,0	Risk of recession in 2018	Stagnation	→
13	INDIA	141	5,5	5,0	4,0	5,0	4,5	Slight economic downturn in 2019	Deflation	↓
13	INDIA	142	5,5	4,0	2,0	3,0	4,0	Sharp economic downturn in 2019	Deflation	↓
13	INDIA	143	5,5	5,0	3,0	-0,9	0,0	Risk of recession in 2019	Deflation	↓

China: most probable one-year scenario

China's housing downturn could continue for several months. Export figures improved in 2014; capital goods prices could go down for months; this could induce exports; growth slightly below 7% is a consensus. On the whole, volatile growth figures could be seen approximately 6–7%. Inflation has been going down in 2014 from 2.5 to 1.6%; this move helped by energy prices could go on in 2015 without being negative; after this energy price downturn, inflation would come back to higher levels.

China: other one-year scenarios

Japan, after 15 years with more than 10% growth, had more than 10 years of 5% growth, could it be the same for China? Not next year! Possibly later; monetary policy becoming more transparent will be a help to foresee growth stabilization. Current account is at very low level; foreign currency reserves are stabilizing at four trillion. So, the yuan could be stabilizing.

China: most probable five-year scenario

Growth could be still slightly above 5% at the end of the five year period; fixed assets growth could go down from 20 to 10%. Export growth could be lower than import growth; inflation rate and current account could be at low but positive levels. Such a change in economics dynamics implies risks and uncertainties, but the Chinese government seems efficient.

Japan: most probable one-year scenario

Scenarios of 2015 are to be built on the number of reforms that will be achieved. Shinzo Abe has been strengthened by the election results. The coalition of the leading party of Mr Abe (LDP) and New Komeito is in good shape and would be able to set the program of revitalization of the private sector by encouraging investment. Labor market reform is essential to shift labor, more easily, from mature industries to growing industries, to give the ability of flexible work time and to pay workers on their results, products or outcomes; a legal framework for a better system to resolve labor conflicts and the participation and advancement of women is also part of this program.

With this program implemented and a cut of 3% of corporate tax, a 2% wage increase is possible and a growth of 1.5% is foreseeable. As the economy has been much weaker than expected after the April consumption tax hike, the next consumption tax hike is not to be seen before April 2017.

Inflation could stabilize at 1% in spite of the yen devaluation. Corporate earnings will be up because restructuring and cost cutting have prevalent effects.

Japan: most probable five-year scenario

Current account is now at low negative level and could remain there; however, Japanese investments abroad are significant and yield more than this current account negative level. Japanese exports would remain at the same level for the five next years and by the end of this period be lower than imports boosted by consumption growth; Abenomics success will likely not be possible; if exports could increase as imports, Abenomics success would very probable; it would be very difficult for the Japanese economy to go out of deflation with only a domestic consumption boost. An export boost is needed too.

Consequently, the Chinese demand for Japanese exports is important for Abenomics success. The Chinese growth has a strong effect on Asian countries: Japan, but moreover Taiwan and South Korea, even Australia.

Asian scenario interaction is an interesting topic for investors as the Asian stock market is not expansive because the Chinese growth deceleration is feared. This fear could be very high so that optimistic Asian country scenarios have to be considered.

Taiwan: most probable one-year scenario

Export demand and domestic demand will be strong again in 2015; the manufacturing investment cycle is in its quick development stage as manufacturing capacity is nearly insufficient in some large groups especially in electronic products. Interest rates were cut to 2%; this level is the lowest since 2006; some inflation could justify some rate hikes: however, it seems that lax monetary policy could last for some time as the Taiwanese dollar has been nearly as strong as US dollar.

Taiwan: probable five-year scenarios

Relationships between the Chinese mainland and Taiwan are problematic; it is not possible to forecast these political and geopolitical developments over five years; Chinese tourists are increasingly numerous; they are able to offset Chinese economy slowdown consequences on

Taiwanese economy growth. In five years, a slower growth is very likely. However, stagnation is a low-probability scenario.

2.6. An efficient one-year scenario generator

As for the five-year scenario generators, the appropriate method consists of using two kinds of generator.

Indeed, a process with two different kinds of scenario seems to be the most appropriate:

– a crisis ESG: for a one-year scenario including crisis, an ESG with a simple structure has to be used. Indeed, it is not necessary to use a really precise modeling in that case. The main characteristics of this kind of scenario are given in Chapter 3 (section 3.1.3);

– a more complex ESG has to be used for a one-year scenario without crisis. This kind of "real-world" ESG has been studied in many papers (e.g. [WOR 11]).

2.6.1. *One-year generator for negative scenarios*

As mentioned above, several studies are quoted in section 3.1.3. The main features of these ESG are the following:

– possibility to integrate a crisis scenario;

– easy to use;

– simple structure.

An example of that kind of ESG is given in [WOR 11].

2.6.2. *One-year generator for positive scenarios*

There are many papers on real-world ESG in particular because of the new insurance reform Solvency II, which requires firms to perform an ORSA and to calculate the solvency capital requirement (SCR). Indeed, the neutral ESG does not allow firms to perform the calculation (see Box 2.2).

The main difficulties of this kind of generator are to calibrate the equation and to choose an appropriate structure.

Characteristics of real-world ESG:

– economically realistic;

– scenarios exhibit the dynamics and inter-relationships;

– used for: capital assessment, balance sheet projection or investment strategic.

Characteristics of risk neutral ESG:

– market-consistent ESG;

– scenarios are consistent with current market prices;

– Used for: liability valuation and pricing.

Box 2.2. *Real world versus risk neutral ESG*

2.6.3. *Other issues*

How can we model with negative one-year scenarios a positive five-year scenario or the reverse?

A long cycle is possible because growth is slow and no excesses will develop in a difficult environment when shock possibilities are able to avoid any euphoria; this kind of cycle cannot be linear, there are different phases and different kinds of shock, much bad news but also positive news. To use a one-year scenario and a five-year scenario is different from modeling, but an autoregressive process with jumps could be compatible with a long cycle. A seasonal autoregressive integrated moving average (SARIMA) process for an inflation/deflation process with the use of hybrid Pareto residue modeling could also fit.

Another kind of cycle could include a major disturbance that has bigger consequences than the jumps of the already quoted autoregressive process; a simple Poisson process for the oil price and/or food price with or without mean reversing process could fit with a benign one-year scenario and a negative five-year scenario.

A Markov switching inflation process could be well fit to a change of period to change the Kondratieff cycle phase [WOR 11]. When the first year is not in the same period as the following years, the chosen one-year scenario is very different from the five-year scenarios, so a Markov switching process could be the right solution.

3

How to Use These Scenarios
for Asset Management?

3.1. Philosophy of equity portfolio optimization

3.1.1. *Optimization and risk-oriented philosophy*

The University of Zurich did some research on features to overcome the weaknesses of the traditional mean variance optimization. Indeed, the mean variance utility function of the Sharpe method is the easiest to implement, but using the second-order moment causes several optimization issues. Variance, as an addition of squared values, does not distinguish positive and negative moves: this is an important problem when there are dissymmetric fat tails. Moreover, in a bear market, volume is decreasing and operators are more nervous; consequently, noise becomes important and adds up in variance. Studies of noise in the markets are numerous: for instance, the papers by Filippi, Lepage, Meyrignac and Pochon on high-frequency statements of stock prices [FRA 12], Kai Yao [YIN 90] on low frequency and Professor Haugen's studies on weekly statements and monthly statements showed important random noises that increase in bear markets. Trading algorithms use filters, such as the Kalman filter for low-frequency information or the particle filter for high-frequency information; smoothing formulas are used by Professor Haugen and many others. Trading systems are issuing more than one order out of two in New York and one out of three in Europe.

Many optimization methods are not valid in a crisis period. For example, methods based on diversification optimization are not valid because minimizing weighted average correlation coefficients is not a good method as these coefficients are very unstable in crisis periods. Diversification quickly decreases in stress periods with increased correlations and fat-tailed distributions.

To overcome this optimization issue, it is interesting to examine optimization working papers (see the bibliography) that include a great many backtests on processes using different kinds of risk measures. Comparing backtests leads to understand how fat tails and risk trend increases can influence the optimization process. In a crisis context, qualities and failures of scenario-based optimizations are important to understand well. For example, an optimization by expected mean return/expected shortfall is very sensitive to distribution fat tails. So, optimization on mean return/expected shortfall can be well fitted to a risk-prone period that can be precisely defined by scenarios.

Optimization based on expected return and a measure of risk implies a choice among the risk measures and the calculation processes; this choice is essential in a century of rising risks; the computation of risks on the past cycles implies systematic underevaluation; optimization using undervalued risks implies too many high beta stocks.

Quantitative asset managers do not generally examine crisis problem risk-oriented philosophy. Consequently, quantitative asset management is, ordinarily, optimized on past variances. This optimization does not take into account fat tail dissymmetry and the rising trend of risk. Compared to an average cycle, next cycle end scenarios could have very different risk levels; so at the end of cycle, many portfolios quantitatively optimized with undervalued risks will have to reduce their risks by sales. When this end of cycle is highly probable, even intuitively managed portfolios will have to sell a lot of high-risk equities; when the next recession appears, as remembrances of 2007–2009 will be very present in the media, a blunt risk aversion increase can be rational or not: that means fund redemptions and big sales of equities which imply higher risk than foreseen. This means that the equities market has a special frailty coming from risk undervaluation; fears of a 2008 renewal could imply an incredible risk aversion, but it is not sure that that fear of illiquid markets could rebound so bluntly inducing widespread systematic risk.

A resilient methodology is necessary because, in a crisis period, the variance-covariance matrix is unstable and correlation coefficients are abruptly moving up at the beginning of the process and are unstable later. So, to optimize, with wisdom, it is necessary to include this frailty in end of cycle scenarios; with a scenario generator, it is possible to avoid all risk undervaluation; thus, the kind of optimization obtained can give a resilient portfolio in case of a crisis.

Considering the risk of misunderstanding and the uncertain understanding of central bankers' explanations and actions, is there an efficient diversification? As central bankers must hide their in-depth convictions, their statements cannot be straightforward and could be obscure. Asset managers have to second-guess, but they could err; therefore, some scenarios are based on various interpretations or possibly misinterpretations. Draghi said on 21 July 2011: "whatever is needed". Markets were not convinced and on 26 July 2012 he said: "whatever it takes [...] markets were convinced that the European Central Bank will act as a lender of last resort, notwithstanding the Bundesbank".

In the face of these uncertainties, diversification is needed, but how efficient is it during a crisis?

Diversification by asset class is not very effective in a crisis period, but diversification based on the study of risk factors for resilient companies, resilient sectors and resilient subsectors is effective. It is the best kind of process for a significant crisis. Risk factors must be updated frequently during a crisis. To manage active exposures to every kind of risk during a crisis process is the best process to allocate funds with a risk minimization aim. Diversification by country was effective in the second half of the 20th Century, but no longer in the Great Recession (2008–2010).

3.1.2. *How and how frequently should we use an economic scenario generator?*

An economic scenario generator (ESG) has to be updated more frequently when the economic situation evolves more quickly; optimization using a scenario generator has to be done more frequently when the market evolves more quickly. To progressively protect a portfolio when a crisis approaches, it is good to modify both the scenario probabilities and the scenario crisis generators.

When a crisis scenario assigned probability has to be more than 25%, it is good to reallocate probabilities. When the crisis scenario assigned probability has to be more than 50%, it is good to reallocate probabilities and take a suitable utility function. When an abnormal crisis comes with more than 75% probability, it is good to take a more specialized economic and/or financial scenario crisis generator with many rebound scenarios, recovery scenarios and convalescence scenarios; however, this generator has to include crisis processes, including social and/or political crisis scenarios and other kinds of crises. This generator has to cover all possibilities, particularly to include scenarios with permanent effects. An asset management using this kind of help could be much better than a portfolio with constant proportion portfolio insurance (CPPI) protection.

When there isn't an approaching crisis, it is necessary, by trials and corrective action, to find the criteria that justify scenario probability updates; portfolio optimization frequency during an ordinary period has to be linked to the kind of asset management and the horizon of the manager; but there are other criteria linked to transaction costs; a record of the costs of portfolio updates after optimization is necessary; a record of simulations done to assess the cost of various update projects could be helpful. Systemizations of these problems are suitable for a homogeneous portfolio with a defensive, resilient allocation of bonds or equities. Some asset portfolios of insurance companies could be sufficiently homogeneous to justify that kind of systemization.

For insurance companies' asset and liability management (ALM), yearly or quarterly use of ESG could induce the same low-frequency use to monitor asset management. Backtests showed that much more frequent use is better. Problems that arise from scenario probabilities need to be updated when recession risk appears to realize. Economic leading indicators are used to indicate as many false alarms or more false alarms as right ones. So probability update is a fund manager's decision as the size change of the resilience pocket; if probability updates are frequently important, it implies a costly portfolio turnover. To reduce this drawback, the manager has to wait to be sufficiently sure about update validity; this cannot be quantified precisely, but it is possible to avoid by back test study pro-cyclical effects which imply excessive turnover.

Markets are now more short term driven than in the past when medium- to long-term management institutions were a stabilizing factor for markets.

Worldwide real-time information, especially for quarterly company earnings, are inducing reassessments. Many regulations are now based on market prices: "mark-to-market valuation" is pro-cyclical and brokers' influence on asset managers is pro-cyclical too. This is destabilizing but some mean reversion could limit some fluctuations. As a result, all processes with jumps are not well fitted to markets. "Mark-to-market valuation" that is now generalized has pro-cyclical effects. Regulations (Basel III for banks and Solvency II for insurance companies) are based on one-year risk evaluations giving one-year scenarios an increased importance.

Institutional mandatory quarterly meetings to appraise management risk/return results for are becoming increasingly usual. All this implies frequent scenario updates and scenario probability updates. In a crisis-prone period, all this short-term monitoring activity makes sense.

3.1.3. *Economic scenario generator for a crisis-prone period*

3.1.3.1. *Simple structure has to be preferred in a crisis-prone period*

In quiet periods, with stable volatility, return distributions of portfolios and even return distributions of many stocks are nearly normal because central limit theorem conditions are not too much of a problem as many stock price drivers can have normal distributions which are not very correlated.

In a crisis period, it is very different; volatility, which can be very strong, increases at the beginning of the crisis process and reduces when the convalescence is clear. Stock price drivers are correlated and their distributions cannot be normal. Consequently, a crisis ESG has to be simple because to model precisely the evolution of a stock is not possible in such a context. Moreover, as mentioned in the following section, jump function has to be included in a crisis ESG: that is why it is no longer necessary to try to model precisely the evolution of stocks.

3.1.3.2. *A few words about optimization in a crisis-prone period*

In a crisis period, optimization cannot be done ignoring fat tails resulting from time-varying volatility. There is no simple possibility to take into account time-varying volatility with time-varying parameters in an optimizing process. A solution to this problem consists of optimizing on skewness or on utility function including skewness. This can be done to obtain a portfolio that is more sensitive to upward moves than to downward moves. To have a portfolio which is resilient to extreme events, it is good to optimize a utility function that includes skewness with a positive factor and conditional value at risk (CVaR) with a negative factor; CVaR is better than value at risk (VaR) because VaR is not subadditive (Artzner coherence); expected shortfall or lower partial moments defined to be sufficiently stable but indicative of bigger risk size could be better than CVaR in some periods. We develop this problem of building a coherent utility function further in this book (see section 3.3).

3.1.3.3. Including jumps function in the ESG

As the size and the timetable of a crisis are unknown, to include flexible jump functions is useful to take into account probabilities applied to the size and timetable that are estimated. Exogenous events with many kinds of scenario can be included in a process modeled by a seasonal autoregressive integrated moving average process (SARIMA) (see Figure 3.1):

– crisis with mean reversing process:

  - without level change,

  - with a foreseeable level change;

– crisis with permanent effect.

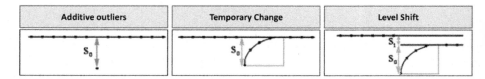

Figure 3.1. *Several kinds of jump function*

As far as core inflation (positive or negative) is concerned, the combination between the Box and Jenkins algorithm to determine the degrees of the SARIMA to be used and the jump functions to simulate the crisis by including jumps – with or without reversing process – is well adapted. Indeed, with fewer than four parameters, their stability is not a problem.

3.1.3.4. *The use of the Pareto law to model residuals*

Insurance companies use Pareto laws to model unforeseeable rare occurrences which characterize the great risks. For scenario crisis generators, it seems well adapted to use the Pareto law to model residuals. Indeed, to model residuals using a normal law is reasonable in a quiet period with moderate cycles whereas in a crisis period, characterized by irrational behaviors and quick risk aversion change, the assumption of normality is inconceivable.

More precisely, to model the residuals of a crisis ESG, the use of a hybrid Pareto distribution is well adapted. This function was introduced by William Gehin [GEH 12] and used by Hervé Fraysse [FRA 12]. Indeed, to build a statistically coherent model, it is interesting to look for an easily implementable distribution better adapted to model the equation residuals, such as a hybrid Pareto distribution.

This distribution is built in the following way (see Figure 3.2): the center is modeled using a normal law and the tails of distribution are modeled using a generalized Pareto distribution. By fixing conditions of continuity, derivability and density, a statistical law depending on four parameters (average, variance and parameters of tail) is obtained.

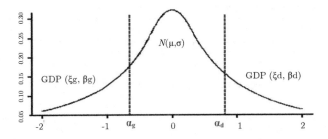

Figure 3.2. *Density of a hybrid Pareto distribution*

It is possible to adjust a hybrid Pareto distribution to a set of data by applying classical methods such as the method of moments and the maximum likelihood. Other more specific methods also exist, such as the Zhang estimator.

The Pareto distribution and the generalized Pareto distribution (GDP) are two different concepts.

The Pareto distribution

The Pareto distribution is a power-law probability distribution used not only for the description of social phenomena but also for actuarial and financial modeling. The cumulative distribution of this law is defined as follows:

$$F_X(x) = \begin{cases} 1 - (x_m/x)^\alpha & if \; x \geq x_m \\ 0 & if \; x < x_m \end{cases}$$

The generalized Pareto distribution

The GPD is a family of continuous probability distributions. It is generally used to model the tails of another distribution. This distribution depends on three parameters: location (μ), scale (σ) and shape (ξ). The cumulative distribution of the GDP is defined as follows:

$$F_{\xi,\mu,\sigma}(x) = \begin{cases} 1 - (1 + \xi(x - \mu)/\sigma)^{-1/\xi} & for \; \xi \neq 0 \\ 1 - e^{-(x-\mu)/\sigma} & for \; \xi = 0 \end{cases}$$

Support:

– x ≥ μ when ξ ≥ 0;

– μ ≤ x ≤ μ - σ/ ξ when ξ<0.

Box 3.1. *Pareto distribution and generalized Pareto distribution*

3.2. Which classic optimization processes are well fitted?

There is a link between rationality and learning: with irrationality, learning is not meaningful. Without learning, what kind of rationality can be seen? The book entitled *This Time Is Different: Eight Centuries of Financial Folly* by Rogoff [ROG 09] showed that the crisis learning process is not so good as classic features are overlooked or disregarded and that original features are overvalued or too well perceived; rationality in a crisis is easier to keep if a crisis learning process gives the ability to compare with many crises and to understand what is similar to the previous crisis.

3.2.1. *Capital asset pricing model*

The capital asset pricing model (CAPM) was improved by Black *et al.* [BLA 72] in 1972; they proposed a method including valuation error management for betas; other explanatory variables (size, growth/value, etc.) were added especially by Fama and French [EUG 92] in 1992 (see Box 3.2 for more details).

Optimization, even if it does not take into account dealing expenses, gives portfolios concentrated on a few lines; these portfolios are very unstable. Some modifications on some input values (such as yields, alphas or errors on alphas or other inputs that can be noise-polluted and that have influence on optimized portfolios) can change the portfolios resulting from optimization outrageously; it is difficult to find reasonable constraints to stabilize these optimization results. So, a quantitative screening on alphas, yields and some other parameters produced by financial analysis can be a good process, especially in a crisis period when stocks are to be judged on resiliency and rebounding capability. There are some hypotheses that are problematic issues for many asset managers:

– normality;

– CAPM-optimized portfolios should be able to borrow and to lend without restriction at a short-term rate;

– CAPM can be generalized by a pricing model with explaining factors.

Arbitrage pricing theory (APT) can cope with numerous factors with a less difficult hypothesis.

In the development of the asset pricing model, it is assumed that:

– All investors are single-period risk-averse utility of terminal wealth maximizers and can choose among portfolios solely on the basis of mean and variance.

– There are no taxes or transaction costs.

– All investors have homogeneous views regarding the parameters of the joint probability distribution of all security returns.

– All investors can borrow and lend at a given riskless rate of interest.

Note that some CAPM hypotheses were relaxed:

– Lintner (72): homogeneous expectations are no longer needed.

– Black (56): risk free asset is no longer necessary to CAPM that is very meaningful since the (2008–2010) Great Recession as treasury short-term rates are no longer risk free.

– Merton (73): continuous time is no longer needed by CAPM; so interest rates stochastic process can be used.

The main result of the CAPM is a statement of the relationship between the expected risk premiums on individual assets and their "systematic risk". The relationship is as follows:

$E(R\_asset) = R\_F + \beta\_asset (E(R\_M) - R\_F)$

where:

– E(Rasset) represents the expected returns of the considered asset;

– E(RM) represents the expected excess returns on a "market portfolio";

– RF represents the riskless rate of interest;

– βasset represents the beta of the asset (i.e. the volatility).

Expected return forecasts are hazardous; factor allocation is more effective than sector allocation and asset class allocation.

Box 3.2. *More details on CAPM*

3.2.2. *Screening or optimization by arbitrage pricing theory*

Screening on many parameters with or without a quantitative formula is a pragmatic process. APT is based on as many factors as it could be good; it could be based on a linear economic model designed to evaluate stock prices or to explain risk. For example, Chen, Roll and Ross [ROS 76] in 1986 used a linear model with inflation rate, interest rates, growth rates and oil prices. APT was first proved by Ross as he found the basic theorem in 1978: if arbitrage factors could not be re-equilibrated, the market will be always inefficient. In ordinary times, when markets are efficient, it is possible to extract arbitrage factors as John Blin showed in [BLI 99]; some funds managed in 1980–1981 with a simple version of APT had very strong returns because they had the oil price among their explaining variables. Asset managers have a hard time when the main drivers of the stock market are changing frequently. Fama and French [EUG 92] showed in 1992 that factors such as the size of a company, value/growth company characteristics and leverage were, when the market is efficient, the right factors to be used in APT. It is better in a crisis, when the market is not efficient, to make allocations after screening stocks on the driving market parameters as a function of the stage of the crisis. Screening can be a dynamic optimization without the heavy process that is difficult to change quickly when a crisis stage changes; screening can be a continuous time process with quick matching to market regime switch; therefore, screening process can be a strong competitor to optimization process. Some comparisons were done by backtesting funds managed with screening processes and ATP managed asset management; look-back bias could be a problem.

The APT assumes that investors believe that the n × 1 vector, r, of the single-period random returns on capital assets the factor model:

$$r = \mu + \beta * f + e$$

where:

− μ is an n × 1 vector;

− β is an n × k matrix;

− f is a k × 1 vector of random variables;

− e is an n × 1 vector of random variables.

APT does not imply hypothesis on utility functions on return probability distributions.

Box 3.3. *Arbitrage pricing theory*

3.2.3. *Black–Litterman (1992) for more stable results*

As an advanced mean variance optimization, investment is done in cash or lowest risk and in the market: in the absence of views that are different from market views, it is a global CAPM equilibrium model. To obtain better returns than the market, asset managers need to bet on scenarios which are well fitted to the reality rather than the market views. The Black–Litterman approach can be seen as a Bayesian approach which assigns probabilities to market views, a probability to *a priori* scenarios of the asset manager that can also have bets on some scenarios based on data; this Bayesian approach stabilizes the portfolio which results from optimization compared to a CAPM-optimized portfolio which is very sensitive to errors and even small changes of inputs.

The methodology, consisting of using assigned weights to both market views and the fund manager's own firm belief, is not well fitted to take into account the numerous crisis scenarios quoted previously. The Black–Litterman methodology with a few scenarios is manageable with a moderate turnover of equities. This could be good if there were a few possible scenarios, a few kinds of crisis process and a few kinds of recovery:

– backtesting done at ENSAE [78] or in ISUP actuarial dissertations showed that comparisons between equity building processes are influenced by recessions and they depend very much on the crisis and the kind of crisis encountered in the period studied;

– to diversify risks with a risk parity process which is resilient in case of recession and crisis is possible and it gives the good results. This process can diversify without overweight on small growth stocks as some other processes do;

– in a quiet period or in a mild cyclical period, if the allocation weights are functions of measures of investment quality (Sharpe ratio example), it could be best with the right rebalancing frequency. However, small growth stocks that are overweighted in these processes are costly at the end of bear markets and very costly if not illiquid in crisis. Classifications and rankings (by Sharpe ratios, for example) are disturbed by crises; recession is usually a bumpy period for these ratio classifications.

3.2.4. *Optimization based on benchmarks*

It is also possible to adopt a passive methodology based on the use of benchmarks.

The main benchmarks are composed of companies that were successful in the past. However, in case of economic paradigm that has changed (as risk trend changed from the 20th Century to the 21st Century), some benchmarks could be misfit. Moreover, market capitalization weighted management cannot be a neutral factor in these cases.

Smart beta management implies active exposure to some factors. This choice of active exposure can increase, for example, small cap exposure or growth company exposure.

3.2.5. *Active management methodology*

A large number of working papers have shown that few managers are persistent outperformers, but there are inefficiencies that can be studied systematically. For example, generalized benchmarking induces value inefficiencies; companies which step in a main index can be bought directly by many asset managers; a price move can be strong. Passive managers and even benchmarked managers do not rely very much on the fundamental financial analysis; therefore, a good, thorough financial analysis based on some 5-year scenarios can find more inefficiencies than in the past when passive management or benchmarked management did not exist. Expected returns based on financial analysis that is based on scenarios and probabilities taken from an economic scenario generator could give a much better optimization result.

3.3. Risk aversion and utility function

3.3.1. *Which risk measures for utility optimization?*

Variance is a good risk measure if distributions are normal. Variance future evolution can be projected. Over a short time period, up to one year horizon, VaR is recognized as the best balance between implementation feasibility for the past and future evaluation; qualities such as longer term risk measurement are not as obvious. VAR can be very unstable and is not possible to project; subadditivity is also a problem for VaR (not coherent with Artzner's criteria [ART 99]); expected shortfall does not have this problem but backtesting requires a much larger amount of data than what is needed for VaR; risk modeling using correlations is not reliable in a crisis. If risk measures can be explained and projected by underlying factors linked by some reliable relationships that hold approximately during the crisis, then it is helpful. These relationships can be copulas, but it is not easy to use them: Kendall and Spearman measures are necessary for testing a copula's ability to describe and model structural dependence. Copula stability is possible but unsure in a crisis period; some backtests showed that copulas were not stable in some kinds of crisis-prone period. It is extremely insufficient to decide which copula would be the best. This kind of optimization is not possible. For utility functions, the use of drawdown is not satisfactory: drawdowns are too hazardous but downside volatility can be used; it is much better than variance. To study fat tails over a period of 5 years is useful to decide which risk measure is well fitted for the problem under review.

With this one-year return forecasted distribution, indicators based on value-at-risk (VaR, CVaR and TVaR) and moments can be calculated using the Monte Carlo method. Five-year scenarios can be used to calculate VaR, CVaR and moments which are used for many kinds of research (see the working paper in the Bibliography) using statistical methods; this research can show which risk measures are best to optimize a portfolio. Different kinds of portfolios and different kinds of period imply different utility functions or a special process.

For portfolio optimization, utility functions are generally calculated as the weighted sum of scenario returns minus a risk aversion coefficient multiplied by a risk measure; variance is the most commonly used risk measure, but it is a good choice only in quiet periods with moderate cyclical moves.

Optimization of many kinds of utility functions has been tried using moments and ratios in the ENSAE working papers listed in the Bibliography (Sharpe ratio, Jensen ratio or Sortino ratio which is better for longer horizon); these utility functions were used in backtests in different periods to understand the qualities and defects of these measures.

Optimization by variance minimization, for example, gives an asset allocation very concentrated on some stocks and also very unstable in difficult periods. Tail risk hedging optimization with risk budget methods is a very costly method up to the time when a crisis occurs. Many scenario generator systems use a variance-covariance matrix (e.g. Ahlgrim, Brennan and Xia) because they assume that the joint law of contingent errors of the model is multi-normal which is acceptable only in "normal" classic economic cycles. These three examples of optimization drawbacks show that the choice of an optimization process is difficult.

When a crisis appears or when a crisis seems extremely probable, the economic scenario probability of the generator has to be updated; normality is no longer found: short-term distributions, e.g. one-year distributions, have *very fat tails*. When a major geopolitical upheaval occurs, such as a war in the Middle East, a change of paradigm could be observed, inducing a change of ESG and a change of portfolio optimization process. It is a change of paradigm that justifies a specialized crisis generator. For this generator, it is necessary to build scenarios for all kinds of crises and all kinds of recoveries. Backtests have shown that this kind of generator has to use, for optimization, a very special utility function including skewness and CVaR as measures of risk and a one-year estimate of a bear market move as an expected return; these backtests showed that this kind of utility function was the best.

John von Neuman and Morgenstein [JOH 44] used an expected utility function as a linear function of event probabilities and scenario probabilities.

Quadratic utility functions can be formulated with an increasing risk factor.

Logarithmic utility functions are convenient. Usually, risk aversion increases when interest rates move up as monetary policy becomes more restrictive.

3.3.2. *Optimization with classic or not so classic measures of risk*

The Sharpe ratio is the most widely used ratio to rank asset management quality measures; studies with backtestings (see the working papers in the Bibliography) showed that this tool, calculated as mentioned on 5-year scenarios, is used extensively in portfolio building optimization and on day-to-day portfolio management. The Sharpe ratio on a one-year horizon can be calculated easily and quickly with one-year return estimates for each scenario; scenario probabilities are used for a weighted average computation and for a variance estimate which has fewer drawbacks on 1 year than on 5 years. Portfolio weights obtained by optimized utility are very sensitive to anticipated returns; optimization gives strong weights to very few stocks that could change very quickly as anticipated return changes somewhat; mistakes are costly. Multivariate volatility modeling or univariate volatility modeling over 5 years is not a resilient method in the case of a crisis; the use of probability-weighted scenarios to compute ratios is acceptable if the universe of possibilities is well covered and "probabilized". Compared to other moments computed on probability-weighted scenarios, skewness is more helpful in crisis-prone periods for any kind of ratio. Sharpe and Markowitz optimizations are strongly influenced by return evaluation which is more unsteady than variance.

Given a utility function u(x), we can define two different kinds of risk aversion.

Absolute risk aversion

The coefficient of an absolute risk aversion is defined as:

$$A(x) = -\frac{u''(x)}{u'(x)}$$

Several classes of risk aversion have been defined, depending on the characteristics of utility function:

– hyperbolic absolute risk aversion (HARA): this is the most general class of utility functions that is usually used in practice;

– constant absolute risk aversion (CARA);

– decreasing/increasing absolute risk aversion (DARA/IARA).

Relative risk aversion

The Arrow–Pratt–De Finetti measure of relative risk aversion (RRA) or coefficient of relative risk aversion is defined as:

$$R(x) = -\frac{c * u''(x)}{u'(x)}$$

If this function is constant, it is called "constant relative risk aversion": CRRA.

Box 3.4. *Risk aversion choices*

It is interesting to compare the mean/semi-variance utility optimization of Markowitz [MAR 52] and that of Neuman–Morgenstern [MOR 47]; semi-variance does not mix positive and negative spreads compared to average as variance does; so a fat tail is not a drawback as in variance; the same can be said for lower partial moments used (in the 1990s) by Fishburn ratios [FIS 82] and by Sortino ratios [SOR 94] for pension fund optimizations; optimization studies were done with backtests in the ENSAE working papers listed in the Bibliography; the results were not very convincing. Markov switching was studied and compared to previously quoted methods in section 3.1.3; this method seems much better, which will be discussed later (section 3.4.1).

Working papers (see the Bibliography) showed how bigger risk aversion coefficients in utility function used for optimization give more resilient portfolios; they also showed that using skewness, and contingently, kurtosis for utility function has been better in crisis periods (co-skewness and co-kurtosis are very helpful even on a short-term basis to manage portfolios, especially hedge fund portfolios); more generally, many papers showed that using, for optimization, some polynomial utility functions based on unusual moments could be a good option as long as scenario generators give *ex ante* precise moment measures.

3.3.3. *Is a polynomial utility an improvement?*

The application of the theory of stochastic dominance (based on the Taylor formula) to the utility function implies that the second-degree term indicates risk aversion by a convexity/concavity indication; the third degree concerns a risk aversion decrease as the portfolio value increases; the fourth degree concerns leptokurtism. For leptokurtic distributions, the fourth order is needed. Kurtosis is not much influenced by noise; as a fourth-order moment, it gives a very big weight to big moves and a very small weight to small moves such as noise; variance as a second-order moment has a noise drawback because it is very much increased by noise; variance compared to kurtosis gives a much bigger weight to small moves, to noises. The fourth-order moment is generally used to study the hedge fund record; higher-order moments can be used to study very speculative option portfolios. So it is reasonable to limit the polynomial utility formula to the fourth order. As a matter of fact, for distributions not very different from normality, Hull and White limit the Corning–Fisher VaR approximation to the third order.

In total, many working papers (see the Bibliography) showed that the use of some polynomial *utility functions based on various moments* for optimization have an added value in difficult periods to manage some funds, core satellite funds and hedge funds. To obtain better optimization, compared to the usual utility function, using variance as measure of risk is a complex process. With the studies mentioned below, more complex utility formulas including VaR and CVaR were tried, but backtestings showed that, generally, moment polynomial formulas give better results, for example (with μ being a risk aversion coefficient):

$$U=M_1- \mu(M_2-M_3+M_4) \text{ or } U=M_1- \mu(M_2- 2M_3+M_4) \text{ or}$$

$$U=M_1- \mu(M_2- M_3+2M_4) \text{ or } U=M_1-\mu(M_2-4\ M_3+M_4) \text{ and so on ...}$$

A solution to settle well-fitted processes, adapted to a crisis period, consists of using the outputs of an economic scenario generator (ESG), on defined periods (this means carrying backtesting) and calculating many ratios or utility functions. For example, in the working papers, the utility function retained is the following: $U=M_1- 0.25\mu(M_2- 2M_3+M_4)$.

The efficient frontier process could be good in quiet periods and better than a process based on ratios. Surprisingly, backtests do not show efficient frontiers reckoned with semi-variance to be very good. After the 2009 crisis, the fund manager's problem was to take an optimization process by taking more or less account of a new crisis; the easiest solution could be to have, in a fund, a specialized resilient pocket; the fund manager has to increase this pocket when a major crisis probability increases. A London business school paper and later a working paper (see the Bibliography) showed that an equal weight portfolio of very resilient stocks is the best allocation in major crisis when risk measures are not as meaningful as usual. This could be very good for the resilient pocket to cope with a crisis; it will be discussed later.

When a crisis becomes probable or foreseeable, a utility function including risk measures seems to be the right solution. Moreover, it seems relevant to use two chosen risk measures: the combination of skewness (with a positive scalar) and CVaR (with a negative scalar) could be the right solution to optimize a portfolio (or the resilient pocket of a portfolio). VaR, which is calculated and used more often than CVaR because it is used by regulations, can replace CVaR; VaR as a quantile is not a risk measure "coherent" with Artzner's criteria, but the CVaR is coherent. Backtests to compare utility functions used to optimize portfolios can show that for some kinds of portfolios a utility formula with CVaR could be very well fitted and it gives somewhat better results.

CVaR measure is more uncertain than skewness and is less uncertain than VaR. Skewness as a third-order moment is, to a large extent, not influenced by noise which grows when liquidity diminishes during bear markets. Skewness as a dissymmetry measure can be very useful in a utility function for strong cyclical markets (to diminish β in a bear move and/or to increase β in a bull move). Co-skewness could be useful to improve the portfolio skewness which is a safety device during crisis.

A working paper (see the Bibliography) on potential maximum loss (using the Pickhands estimate or Hill estimate) gives a comparison process for optimization (especially CVaR- or TVaR-based optimizations). These optimizations could be too linked to worst case scenarios even when a crisis is unsure. Finally, in these situations, the soundest process would be to increase progressively a resilience pocket allocation and to use, for the remaining funds, an optimization with a mean/VaR and skewness utility function formula ($U = M1 + \mu\, M3 - \pi\, VaR$); it is difficult to do better than to backtest on the period 2004–2006 and in the 1990s to see how secure this process is. The resilient pocket allocates the same weight to some equities of the best-known resilient companies with as much sector diversification as possible. To replace in a utility function, simply, variance by skewness gives resilient portfolios, but not very dynamic ones as backtestings showed.

3.4. Better fit processes for a crisis

3.4.1. *Markov regime switching optimization best fit for a crisis*

At the beginning of a crisis, it is necessary to take account of a complete change of environment that implies fears and several or many irrational behaviors (especially if a catastrophic worst-case scenario is to be feared). A regime switching from a classic economic and financial scenario generator to a crisis scenario generator could be defined as a simple Markov chain. The transition to crisis state has to be characterized by a probability of switching from a scenario with leading indicators showing a possible crisis to the obvious beginning of a crisis; which crisis? Many crisis scenarios are possible at this stage. The beginning crisis can develop with or without contagion, with or without opportunistic reflationary government policy and so on. Each crisis scenario has to be built with a development, a mature pattern and a way out to be built up to an end of convalescence state when a normal generator can fit again well.

An asset management process can be defined by switching from ordinary ESG to crisis ESG when the asset manager assigns to crisis scenarios a sum of probabilities higher than 75%; the reverse switching could be done when the way out of crisis would have a 75% probability in the asset manager's mind. A high probability is needed because this ESG switching to a very different crisis ESG implies a very important portfolio shift that is costly. Despite this important switching cost, it is necessary to compare its own probability and portfolio diversification to those of asset managers betting in convalescence, to be aware of market views. The 2009 bull market showed how these bets could reverse the bear trend and help re-establish some trust.

Malcolm Kemp, in his book *Extreme Events* [KEM 10], takes an example of a regime switching model with a normal regime and a "bear" regime that is not much different from our problem from normal ESG to a crisis ESG. As a result, equity markets are much more correlated on the downside than on the upside; this asymmetric correlation is a much larger phenomenon for a crisis than for an ordinary cyclical downswing. For example, the crisis beginning in 2007–2008 involved a general downtrend of all markets with strong volatility; many bond markets and other markets linked to rates became illiquid as trust was going down.

As mentioned previously in this book, it is clear that the time series behaviors of financial variables have several different patterns over time. Consequently, using a Markov switching model, based on the combination of two or three dynamic models, can be an effective means to model the evolution of economic variables.

Below, we describe a simple model built with two autoregressive models, given the value of an unobservable state variable "c_t":

$$A(t) = \begin{cases} \alpha_0 + \beta * A_{t-1} + \varepsilon_t & if \ c_t = 0 \\ \alpha_0 + \alpha_1 + \beta * A_{t-1} + \varepsilon_t & if \ c_t = 1 \end{cases}$$

The model described above is a really simple model useful to describe the concept of such a theory. However, of course, the equations and the condition of switching can be more complex in order to be more accurate to the economic phenomena we want to describe.

For example, the condition of switching can be calibrated with the evolution of leading indicators.

Box 3.5. *Markov switching regime*

Kemp quoted Ang and Bekaert [ANG 02, ANG 04] concerning arguments in favor of regime switching models applied to cyclical market trends in the USA and the UK. For example, asymmetric bivariate generalized autoregressive conditional heteroskedasticity (GARCH) model that could seem to be a well-fitted concept to cyclical equity markets correlation pattern is not as good as a regime switching as defined previously. The authors explained this better fit because the "RS model" was able to capture persistency in both (conditional) means and second moments; they also explained that RS models are more suitable than traditional mean variance optimization for investors who use skewness to improve the portfolio resiliency. When a manager looks for a crisis, skewness and co-skewness are helpful to monitor and increase the portfolio resiliency. More generally, the authors concluded that the RS model is good "for investors with preferences involving higher moments of returns". Many hedge fund managers have a preference for kurtosis as a measure of risk; noises have bigger effects on variance than on kurtosis, as mentioned previously. Downside risk statistics such as lower partial moments or semi-standard deviation could be useful as complementary risk measures in a crisis-prone period with a very hazardous downside risk.

Regime switching models imply a difficult optimization; Ang and Bekaert [ANG 02] implemented a process with numerical integrations; numerical integration of each scenario is used to compute the probability weighted mean and risk measures.

3.4.2. *What could be said about cost when the method is changed?*

The process to change the optimization method and/or probabilities of scenarios when a crisis is perceived involves a big and costly portfolio change. In the case of an abnormal crisis process, a depression process perceived as a change of paradigm justifies a costly portfolio change; it is necessary. A special generator enclosing many crisis process scenarios, all possible scenarios if possible, is needed; some of these scenarios take account of very large debt repayments in private sectors, going up to the debt cutting obsession as in Japan. As the dynamics of deflation are different from the dynamics of inflation, the Markov switching for non-stationary time series is well adapted to switch from an autoregressive process for inflation to another one for deflation.

The Markov switching method is well adapted to these changing environments; backtests seem to indicate that it is less costly than the previously quoted change of method. The Markov switching method is well adapted to shocks with a change of autoregressive process to take account of a food price shock or/and oil price shock, for example. It can be used without difficulties with three possible switches and with a greater number of possible switches. Cost of transactions needed to build the after switch optimized portfolio is a problem. This cost of changing a process is conditional on the previous state of the system; a transition matrix between these two successive portfolio states needs to be estimated.

It is less costly to change some scenarios and some probabilities; various process comparisons were done with backtestings reckoned by working groups (see the Bibliography); however, the Markov switching for non-stationary time series is well adapted in spite of higher cost. For example, switching from an inflation autoregressive process to a deflation autoregressive process has to take into account the private sector debt repayment process in a deflation period. In a low inflation period, profit maximization is still in force, but, in a deflation period, debt repayment is an increasing motivation (as seen in Japan). The search for maximum diversification involves different asset allocations in different kinds of economic situation; in a deflation era, maximum diversification is rather different. For this kind of volatility reducing policy, Markov switching can be useful and not too costly.

The Markov switching method can also be used with three or four autoregressive processes to take account of inflation shocks (wars, climate shock, food price shock and/or oil price shock). Another way to model these shocks is a simple or a composed Poisson process with small and fixed probabilities. Markov switching is better especially if probability shock is not so small and if inflation shock is significant and lasting (such as war in Vietnam which is not an oil producing country).

Direct or indirect tail hedging is also a volatility reducing policy that can include Markov switching usefully; cost is very difficult to evaluate, to plan and to monitor. Cost is lower and easier to monitor when a generator change is applied to simple methods for which commentaries are given below.

The use of a scenario generator, even a small one, can help risk parity equity portfolio building; Markov switching could help to have a good risk diversification after a period change; risk parity can be made stock by stock or industry by industry or by other categories. Diversifying, as there are some categories of stocks to manage Markov switching, could help:

– in an equity portfolio building with minimum variance process, diversification is a problem; Markov switching between two small generators can be used to determine in particular the cash proportion, risk diversification and transaction cost that will be moderate;

– in a core satellite approach, portfolio rebalancing can be improved by using risk measures evaluated on "satellites' categories". A small generator and a Markov switching can help rebalancing with moderate costs.

3.4.3. Risk diversification: risk parity or risk budgeting?

3.4.3.1. *Risk parity*

Risk parity is now fashionable among institutions but is criticized because it is a pro-cyclical kind of management which is contrary to the value-driven process: the contrarian process which results in an increased exposure in difficult times and a lower exposure in euphoric markets. Risk parity can be applied to defensive sectors and to resilient companies without pro-cyclical inconvenience. Often, *risk parity* strictly applied with every line of the portfolio at the same level of risk as other lines is very costly; as risk measure volatility is important, frequent rebalancing is costly. Risk parity gives to portfolio a much more stable global risk level if reallocation is frequent that all the methods previously commented, but portfolio risk stability is obtained at a high cost. Portfolio turnover has a cost function of reallocation frequency. Risk parity can be done at the level of every line of the portfolio, but turnover will be less costly if risk parity can be done by industry, by country or by other asset class categories. Risk parity has advantages in many kinds of process, but the choice of a risk measure is not easy because rebalancing the portfolio cost is not obvious. In a crisis, the effects of increased correlation coefficients on portfolio risk are significant; these coefficients' instability in a crisis is rather problematic. Risk parity is an elegant method, particularly among pension funds which have a very long-term kind of asset management. They can avoid frequent rebalancing.

The risk parity method endeavors to give a weight to different categories so that the risk of the investment in each of these categories is equal or analogous. The problem resulting from the correlation coefficients' instability during the crisis means that this parity could be illusive, especially during the crisis. Which risk measures are to be used? It is convenient to manage a risk parity process precisely based on skewness or on expected shortfall risk measures that are more stable than VaR as a measure of crisis risk. Comparisons were done in the ENSAE working papers listed in the Bibliography:

– Some institutions work with the marginal risk parity concept which makes sense if and only if *ex ante* risk measures are as precise as possible with a scenario generator. A marginal parity could be a useful concept to maximize the global skewness which is rather stable with skewness parity for each category.

– Using every crisis scenario, the financial study of each company that could enter into the portfolio gives an idea of its resilience and of the kind of correlation and copula that could be seen in these scenarios. A judgmental conclusion for each company and each category leads to a judgmental equilibrium of risk that can give a portfolio not too far from risk parity and rather good diversification in a crisis period.

3.4.3.2. *Risk budgeting*

Risk budgeting is a useful concept for asset liability management of financial institutions. This method endeavors to allocate a predetermined weight to each stock category so that reallocations keep the risk level of these categories nearly equal to the budget decided by the manager; therefore, a scenario generator is very useful. The aggregation problem resulting from the correlation coefficients' instability during a crisis means that this budgeting could be illusory if not frequently rebalanced; rebalancing of the portfolio could be costly in an unstable environment, unstable correlation coefficients, with bad visibility and imperfect knowledge and understanding of what is happening. Risk budgeting can be used to diversify better in particular. If risk budgeting is defined on large categories, rebalancing would not be too costly; as a consequence, the concept of "risk budgeting" is really useful for financial institutions' asset liability management as well as for fund asset management. As so many crisis risks and so many kinds of market movements are possible with a very uncertain foreseeing ability, a simple budget allocation could be done with:

– a strategic portfolio budget, with quality equities that are resilient so that market fluctuations resulting from a crisis or cycle would be bearable;

– during crises, a tactical portfolio budget, designed to be flexible with liquid equities which stayed liquid during a crisis and able to quickly rebound, would be bearable.

This kind of simple risk budgeting is a process without costly frequent rebalancing. Risk budgeting can aim to give a weight to different categories so that the risk of the investment of these categories would be equal to the budget decided by the manager; this would incur more transaction costs.

3.4.4. *Minimum variance policy*

This policy limits investment to very few sectors and companies; these few companies are expensive when *minimum variance* is fashionable. The 500 least volatile ES equities have an average volatility of 10%. A minimum variance fund compared to a big cap fund has a small decrease of volatility and less diversification. Diversification is the minimum variance policy problem which is increased when the most diversifying stocks cannot be very overweighed because risk parity is applied. A working paper (see Bibliography) shows that in the last part of the 20th Century, equal weighted portfolio with minimum variance was underperforming the minimum variance efficient portfolio; in that period of bull market, this result can be explained easily; in a bear market, it is different.

Minimum variance gives big weight to four very conservative sectors that are correlated between themselves in classic cyclical movements and big weight to most conservative companies that could be good investments but not so diversifying. If constraints are added, this drawback is increased. In the latter part of a bear market, institutions needing cash sell the conservative equities, whose stock prices diminish the least. The lack of diversification of minimum variance portfolio at that time is harmful:

– total risk contains a lot of correlation-linked risks; as correlations are already high, their increase to one in crisis when fear increases correlation abruptly is not big; so variance could be minimum in an ordinary economy and rather moderate in a crisis economy;

– minimum variance is rather popular, especially among the US institutions because it is such a large market that there are more than one hundred resilient stocks. In small markets, minimum variance means insufficient diversification.

The 500 least volatile ES equities have an average volatility of 10%. The 1,000 biggest have an average volatility of 12%. The 1,000 most volatile have an average volatility of 20%.

A minimum variance fund compared to a big cap fund has a small decrease of volatility and less diversification. Risk parity by comparison has a very large investment universe; the choice of a risk measure to compare both methods is obviously varied. Comparisons of the Sharpe ratios of these two methods make sense.

3.4.5. *Resilient equity portfolio construction with same weight stock allocation*

Same weight allocation is interesting, particularly for managing resilient portfolios or resilient portfolio pockets. Last, but perhaps not least reference, a portfolio building devoted to crisis protection was studied at London Business School and later by ENSAE students (see working papers in the Bibliography); it is obviously the simplest process to implement. When using it only for the most resilient sector and companies screened by thorough financial analysis for 15 years, it gives the same weight to several very resilient companies trying to obtain the best diversification in this simple allocation. This method can be compared to the risk parity method with low beta preference that is used by many German institutions. Backtests showed that during the crisis-prone 21st Century, both methods had good Sharpe ratios, but the same weight stock allocation does not mean rebalancing frequently to keep these weights equal; rebalancing at each stage of the crisis is sufficient; it did not incur high turnover costs. To diversify a *same weight allocation* better, it is obviously possible to make a same weight allocation in a two-pocket portfolio with the big stock pocket double weighted compared to the resilient middle-sized companies' allocation. Diversification by size more generally makes sense in bear markets and, moreover, in a crisis.

There are also companies that are resilient in a crisis because they are research leaders with an ability with innovative products well adapted to the market mood that give them a solid tactical situation in all scenarios; those kinds of companies have low leverage because they have high and rather stable margins. These kinds of companies can be the long-term investment of a strategic pocket with a long-term horizon of institutional portfolio management. These kinds of resilient company can be found by financial analysis of companies' cash flow, earnings and dividend studied over the long haul to have company resilience proof:

– in scenarios of shock, crisis and depression on which CVaR and VaR can be evaluated;

– in scenarios of inflation and deflation.

Investments in this kind of resilient company could be suitable for many asset managers, particularly for those who choose a core satellite approach; a satellite specialized for this kind of long-term horizon for companies that are creative and resilient makes sense not only in a crisis, but a resilient conservative satellite would be good in crisis scenarios; a generator is effective to compute many risk measures for each satellite and globally for the core satellite fund: studies on tail risk and tail dependence of satellites help rebalance risk by satellite manager switching in a process that aims to increase resilience (or Sharpe ratios). Tail dependence measured by copulas could be an improvement.

Correlation parity between satellites and/or between satellites and core is a policy used by few institutions; as correlation coefficients are unsteady in unstable periods and converge quickly to one in a crisis, this policy seems understandable only in quiet periods.

When conditions are very unsure and when a crisis-prone period seems to be on the horizon, the better approach to build a resilient portfolio – on the go – consists of investing in resilient company equities with the same weight allocation; in this case, scenarios are only useful to analyze companies and forecast their earnings for different kinds of crisis; to decide that a company is resilient to a crisis, it is better to consider many crisis scenarios and recovery scenarios.

Backtests computed and studied in the ENSAE working papers give ideas to build a well-fitted asset building process; but, all this cannot be valid if these measures of risk are not well fitted to the future; so, an optimization process cannot be good if risk measures are not well computed on a good scenario generator, an ordinary one or a crisis generator depending on the kind of period that the manager has in mind. Backtest experience provides a main idea and a general understanding, but many complementary backtests are necessary to confirm the main idea and to build a robust process.

3.5. Crisis process for equity portfolio optimization

3.5.1. *What kind of generator to optimize?*

As correlations are very unstable in a crisis-prone period, the Wilkie ESG concept seems better than Brennan and Xia's ESG concept. Inflation rate, positive or negative, is the only factor of crisis spread to markets in these ESG models. A European cold war without big defense budget increase can involve dangerous geopolitical trends over the medium and/or long term without any significant effect on inflation rate, but this kind of geopolitical trend can increase risk aversion and be negative for some markets. The cold war in Asia is already at the stage when China and Japan are increasing their defense budgets quickly. Japan is still far from inflation and the Chinese economy is decelerating; nevertheless, an arms race over the medium term could mean some rebound of inflation. The choice of generator concerns experts' opinions and ideas on the future.

Wilkie ESG with monthly and seasonally adjusted data is based on an inflation process (inflation rate of a year is determined as a function of the positive or negative inflation rate of the previous year). Equities are evaluated with the Merton model with jumps; bonds are evaluated with the composite regressive model. This generator was studied by Anquetil *et al.*

As the Wilkie ESG does not include geopolitical risks or sociopolitical risks, Huchet, Mikaelian and Revelle added a Poisson process with a strong oil shock (1% probability per year seems low now; this paper was written in 2011); they added with the same probability a strong agricultural raw materials shock able to imply a sociopolitical crisis; they also added smaller shocks with a shorter mean reversing process.

These ESGs have integrated market models based on an autoregressive process for inflation rate which determines rates and dividend yields; so, classic models can be used for equity markets, bond markets and real-estate markets. Experience has shown that ESGs are more resilient when they have a few parameters. For inflation, Vasicek's model with a mean reversing process is the most usual basis for ESG; for equities and real estate, Black and Scholes's model is the most widely used method. For bonds, Hull and White's two-factor model or the generalized Vasicek the model with two factors is generally estimated.

Resilient equity earnings visibility implies, for each scenario, a possible modeling with few explanatory factors; these factors are helpful to determine criteria on future earnings to work out optimizations and probability updates.

3.5.2. *Optimization with risk budgets for crisis resilience control*

Many kinds of crisis are possible, particularly over a five-year horizon. Consequently, an evaluation has to be made on the main risk categories: geopolitical, monetary, banking, financial and economic; sociopolitical crisis is contingent on the previous crisis process. Risks induced by each scenario produced by the generator are to be classified into these categories and added together. This kind of aggregation has to be done with correlations being equal to one because a crisis will push most of them to one. So, this computation overestimates the total risk. Total risk valuation by kind of crisis helps to decide which kind of risk has to be diminished and how to do it. Over a one-year horizon, it is easier, and the number of scenarios is smaller because many kinds of crisis have a negligible probability and scenarios can be more accurate. A portfolio allocation has to be made between:

– a strategic portfolio built to be resilient over a five-year horizon with a risk budget constraint, built with quality equities that are resilient so that big economic and financial fluctuations resulting from the crisis have small probabilities to erase their resilient company status. Some risk budget constraints can be put on some sectors or categories;

– a tactical portfolio designed, with a risk budget constraint, over a one-year horizon, to be flexible with liquid equities which stayed liquid during the previous crisis and which showed quick rebounds.

This is a simple and clear process of risk budgeting; many more complex risk budgeting processes are possible to build resilient portfolios. Risk budgeting with CVaR constraint seems the best; some backtesting on the 1987 crash is helpful to define the risk budgets.

3.5.3. *Insightful comparisons between optimized portfolios*

3.5.3.1. *Stress tests and reverse stress tests*

There are scenarios with probability deemed minimal or even negligible: black swan scenarios and hypothetical scenario "Bayesian priors" that are not included in the previous process because they are not considered as possible. Portfolio valuation for these scenarios which imply extreme stress is interesting to study fat tails and extreme CVaR and TVaR that would result of very unlikely stress; correlation between fat tail extreme ends of stocks included in a portfolio is interesting; however, this kind of correlation is also interesting for stocks that could be included in portfolio. The reverse study in the case of stress scenario loss is interesting too: how a portfolio loss would be allotted between these portfolio stocks if correlation is equal to one or if correlation coefficient is equal to correlation coefficients quoted previously. These studies induce insightful thinking on weaknesses and help to find the box scenarios that do not usually come to the mind as pervasive contagious crisis and worldwide systemic crisis.

3.5.3.2. *Optimization comparisons and reverse optimization can give insightful views*

First step: is there a scenario that would induce a portfolio analogous to the optimized portfolio obtained by the process chosen by the asset manager? Other steps could be useful: comparisons between portfolios obtained by another optimization processes are interesting; if possible, three other optimization process comparisons are very interesting: classic mean variance optimization and two non-classic return risk optimization with a utility function used in the ENSAE working papers: mean/skewness or a more complex utility function with mean, skewness and CVaR or with mean, skewness and value at risk; comparisons between portfolios obtained by many other optimization processes are too time-consuming; therefore, they are not done.

Optimization done with the process chosen by the asset manager can be reversed to obtain insightful views: it is possible to work out a scenario that gives the same alphas as the optimized portfolio implies; the smallest alpha among these implied alphas shows which line of the portfolio is to be sold first. Backtestings of that kind can be done in the previous crisis period. So, it is possible to see how well a risk return model optimization model would have performed in the past crisis and how well problem lines could appear in a reverse optimization.

Other kinds of optimization can be studied by the previously described process: for example, minimum variance. How well will this defensive risk model perform in a crisis? Has it already been applied in past crises? Which kind of low beta stocks could be problematic? When an optimized minimum variance portfolio is obtained, it is also a very interesting exercise to research if there is a scenario which could justify an analogous portfolio to the portfolio obtained by optimization on the universe of deed possible scenarios with assigned probabilities.

3.5.3.3. *Stability comparisons and complexity drawbacks*

With the variance-covariance matrix unstable, during a crisis and even some time before a crisis, computations (for example, risk aggregation,) on risk are not meaningful; but, risk measurements based on scenarios are always meaningful; not only meaningful, but insightful when comparisons on various optimization method results and/or comparisons on reverse optimization method results are studied. However, probability-weighted scenario computations are a Bayesian approach to optimization which has a drawback: probabilities assigned to scenarios, even if they are by experts, are subjective. To try to build a scenario for every possibility is a subjective process; for example, the Eurozone crisis could generate a great number of scenarios with or without the transformation or disappearance of the Euro; this kind of problem could be analyzed by an incomplete information game theory, but the resulting batch of scenarios would be complex. Stability and complexity imply important drawbacks because optimization results are very sensitive to small input changes of probabilities and/or to scenario change, when stability is down and complexity is increasing.

Multi-period optimization would be very useful but it means many more parameters and more complex processes; this implies less parameter stability. Introducing transaction costs increases the process complexity; there is not a recognized method to cope with transaction costs including liquidity cost; this is an important problem during a crisis; therefore, endeavors to better grasp managerial problems imply that the optimal portfolio could be very sensitive to inputs, assumptions, evaluation errors and insignificant market changes. To limit instability problems, it is possible to put constraints on the optimal portfolio by risk budgeting. To divide a portfolio as mentioned into two risk budgets – a resilient pocket and a classic pocket – is the simplest risk budgeting method.

3.5.3.4. *Modelization and backtestings*

Modelization of risk measures with explanatory variables such as interest rates, productivity, growth, liquidity and earnings is useful, especially to study resilient and not so resilient stocks thoroughly. A choice has to be made from amongst as many as 50 variables to forecast risk valuations. Fixed parameter modeling of equities can be criticized but quite simple variable relationships can easily be modified make them more complete if necessary. So, resilient companies can be better selected by taking into account market rational or irrational factors seen in the previous crisis.

Backtestings with scenarios of past crises show the limits of this kind of forecasting and give an order of magnitude of expenses to modify and diminish risks. ESG can be very useful at this stage of risk cutting with or without an optimization process.

Risk aversion moves up as the emotion of crisis recognition rises sharply in the early crisis stage; risk aversion remains high in bear markets that make risk cutting very costly. Strong risk aversion leads to market illiquidity and a vicious circle of disorders that increase fear and risk aversion: risk aversion increase is a vicious circle factor of risk increase. Emotion is a major factor inducing irrational behaviors that increase crisis. At this stage, it is too late for risk cutting.

Scenario generators are used by institutions but not so much by asset management companies, even if it is a methodology that very much improves risk measures and risk management in a crisis process. The use of some scenarios, in order to forecast earnings over a longer time span than the horizon usually used by financial analysts, could give the ability to measure risk without being influenced by the evolution of the stock exchange; this is very interesting in a crisis period:

– a black butterfly crisis means that a reversal process can be modeled using a Markovian process, autoregressive process with jumps; so, the expenses to increase risks again can be valued and taken into account when cutting risk appears good;

– a black swan crisis can be modeled as casualties or as a casualty process that could be an autocorrelated or purely random Poisson process. A casualty in a high-risk environment can be treated with disastrous casualty insurance methods;

– computations of risks based on the inflation model by Ornstein–Uhlenbeck and the bond prices model by Hull and White allow rational choices to increase or diminish risk; the cost of this move can be estimated and compared to the expenses budget aimed for. This cost can be improved by trials and correcting actions.

The best process to use an expenses budget is a pragmatic one: it is to establish a list of the best stocks to increase utility function of a portfolio with the utility increase amount and the estimated cost of the purchase. A list of potential sales by the portfolio with utility decrease amount of the line sale with estimated cost of sale has to be established. An optimization utility process has to be carried out under the constraint of the total budget cost of arbitrages. This can be done simply step-by-step. The first step is the portfolio line with minimum utility decrease is the first sale used to buy the best stock; this simplest process which is not as good as the global optimization has to continue up to the time when the total cost of realized transactions is equal to the defined transactions cost budget.

3.5.4. *ESG and stock management*

Crisis ESG is needed to manage stock resilience portfolios.

3.5.4.1. *Quick historic*

The real estate bubble burst by mid-2007; the credit bubble could be compared to a crisis which reduces by half savings and loans, 20 years earlier; this kind of "black butterfly" crisis with above average recessions could be forecast early with a good probability of a right forecast; but, the 2007 crisis was another kind of crisis: a "black swan" crisis became obvious when money markets became obvious; there was a problem with forecasting this crisis: the published figures about real estate credit were wrong (conduits and derivatives). The International Monetary Fund (IMF) published much higher figures in 2008; crisis scenarios could not have adequate probabilities; Roubini and a very few economists spoke about crisis-increasing probabilities. When UBS bought Californian real estate institutions nearly insolvent, the number of economists who wrote about a crisis increased quickly; when the Northern Rock bank run was stopped by the British Treasury, an international crisis was obvious; Bear Stearns became insolvent, and a "black swan" scenario was very likely. In early August 2007, a liquidity injection announcement seemed a very hazardous process to increase the Federal bank system's total balance sheet, threefold; the European central bank was just as quick in summer 2007 and other central banks increased their balance sheets much more than in the past. Many scenarios could be made on the result of this exceptional monetary process, but, bank treasurers were the first kind of operators to understand the crisis process; from 7 August 2007, abnormal spreads (short OIS swap rate-deposit rates, for example, moving up to early August as some BNP Paribas faced valuation problems; but the TED spread gave the first signal in June when some Bear Stearns funds had some problems) appeared showing treasurers' protective moves. On the whole, a crisis scenario generator including many kinds and sizes of crisis was needed to give a probability to the Roubini scenario which was published in September 2006. The Japanese kind of deflation and some kind of depression could have some probabilities when the Bank of Japan decided in 1989 on a monetary policy to burst the global bubble. A switch to a crisis economic scenario generator makes sense when central banks announce bold moves.

The result of these central banks' risky moves was a success: to diminish the probability of depression: they replaced the worst-case scenario by the Great Recession in 2007-2010. The Bank of Japan's reflation moves linked to the budget reflation plan were not successful because taxes were increased too early with a recession as a result; certainly, the Japanese deflation experience and the USA experience of the 1930s were very much in Bernanke's mind for years and in central bankers' culture; however, the central banks' extraordinary programs with very low rates were risky leverage for them. When such risks are taken by central banks, a crisis economic scenario generator is needed; perhaps, a crisis economic and financial scenario generator would be better but more numerous parameters are needed.

3.5.4.2. *How to react during the next crisis?*

In the next crisis, central bank leverage will still be a problem because it is not possible to multiply again by two or three the central bank balance sheet total. Nevertheless, it was necessary for Japan to face a huge debt repayment process in a situation which very much diminishes their monetary policy clout. Perhaps, the Japanese monetary policy is no longer efficient.

What can be done in the case of a multi-crisis process? In this case of deflation process, the private sector debt repayment process and interest rates at near zero (as in Japan) would make monetary policy totally inefficient. This scenario has a significant probability. Excessive monetary growth implies many kinds of financial bubbles and disorders. The banking system's increasing complexity implies that improvement of bank monitoring will always be late and insufficient and that systemic risk will remain. Public sector refinance will be very difficult in the case of a new crisis as a reflation program will again very much increase debts to the gross national product (GNP) ratio. All these risks have to be monitored and aggregated often and quickly and precisely with economic and financial scenario generators. If many different scenarios on these matters are pessimistic, there are more optimistic scenarios as the central bank seems to learn quickly; a crisis scenario generator has to include optimistic scenarios justified by the central bank learning.

When risks are changing quickly in a crisis process, a well-trained management team is able to update many scenarios, and to study contingent probability change and change the optimization process. Well-organized and well-trained teams will be able to stay rational when the atmosphere is as disturbing as in September 2008 with the Lehman failure; when the tone of the markets and trading rooms are deleterious and brokers' advice pernicious. Before all kinds of crisis and all kinds of catastrophic news with unforeseeable consequences, crisis generator methodology of quick but rational updates of all risks with a resilient aggregation methodology of risks is an ace to outperform competitors. Difficult times and difficult markets when so many behaviors are irrational could be the best opportunity to outperform as a crisis scenario generator is a tool to stay rational. Optimization process rationality could be an important help to outperform in fearful periods.

3.5.4.3. A crisis generator is needed

During the Great Recession, with the Lehman failure in September 2008, most quantitative asset management in equities had performances because models were based on economic and financial explanatory factors which were good before the crisis but not at all during the crisis process and the recovery; the quantitative asset management teams which changed their models quickly were able to do better; with crisis generators, they could have outperformed strongly. Even in every catastrophic piece of news, war, trade war, currency war and end of the Euro, there is an opportunity to outperform using economic and financial scenario generators. There are also optimistic scenarios with hope that the G20 governance becomes more effective to stop wars and to cut deals even on international monetary problems that have not even been mentioned up to now in G20 meetings. The worst case is never certain, central banks are learning to avoid worst-case scenarios; there are always best cases that economic and financial scenario generators are able to remind users. An optimization process cannot only be mesmerized by worst-case scenarios; it could help outperform very early in recoveries when contrarians bet on a global convalescence.

When a crisis is perceived with the feeling that this time is different, the kind and the size of the coming crisis are not possible to judge; but it is the right moment to use a crisis generator of scenarios and to give probabilities to scenarios modeling different kinds of crisis, different sizes of crisis and different kinds of crisis processes; this kind of generator must be prepared in advance because it is too late to undertake that kind of work when a crisis is coming. By the end of the crisis, a more classic generator is necessary; later, the change of probabilities of the various scenarios is needed as the recovery situation becomes clearer and clearer. When recovery leads to a normal cycle, a generator is less useful if there are no geopolitical risks, climatic risks, sociopolitical risks or even epidemics. All contagious risks cannot be precisely evaluated without a generator. With crisis scenario generators, it is possible to outperform the competition because, in difficult periods, classic quantitative approaches, disturbed by a lack of rationality and by fear, have a difficult time; better scenarios and best scenarios can help to outperform bull markets as well; an optimization process, based on rational measures of risks including worst case scenarios, helps to stay rational in bull markets and even in euphoric markets.

3.5.4.4. *Example of crisis generator*

It is really difficult to find studies on crisis in the literature. Indeed, this topic is not currently addressed in most of the insurance companies or by fund managers. As mentioned earlier, the main reason is that the fund manager has not yet taken into account the increase of major crisis probability seen since the beginning of the 21st Century.

The main example of the construction of a crisis generator is found in a dissertation from ISUP 2012 by Hervé Fraysse. The structure retained is the Wilkie structure (1985). The choice of the Wilkie ESG structure (1985) results from several elements:

– in periods of crisis, it is better to choose cascading structures because they are more stable than the correlation structures. In addition, a major advantage of these structures is that they are more flexible: possibility of adapting the ESG while keeping the same structure;

– simple to use: the equations of the model are simple – no mathematical complexity;

– this ESG model has fewer parameters than the Wilkie 1995 model: a retrospective study was realized on the Wilkie model to test its robustness. It concludes that the Wilkie model is more successful in its first version (1985) than in its second version (1995);

– this ESG model is a good balance between statistical complexity and speed of simulation. Let us recall that the complexity of the statistical extensions improves the adequacy to the past without inevitably improving the quality of the forecast;

– allowing projections over 1 or 5 years.

The diagram below illustrates the ESG structure.

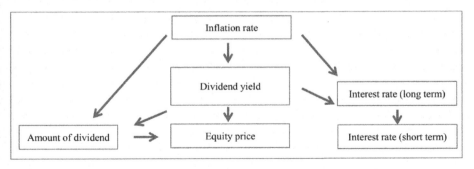

Wilkie's ESG structure

The plentiful literature on this model allows us to avoid some stumbling blocks and allows this model to evolve. The critics are essentially concerned with the specification and the quality of the adequacy of the model. Several English actuaries strongly criticized the model, in particular Kitts, Huber and Geoghegan *et al.* [ANG 04]. The critics are mainly concerned with two elements:

– the model specification and the initial parameter dependency: Huber (1995) underlines that the overparameterization of the model implies a strong dependency of the results on the initial parameters. The equations contain a mean reverting process and the choice of the mean conditions the future results and the projections. Consequently, specification choices must be made with caution;

– the biases linked to the estimation: Kitts (1990) shows that the series of inflation, initially used by Wilkie, is not a stationary one. Therefore, he criticizes the modeling of the inflation by a one-order autoregressive process. Geoghegan *et al.* (1992) show that the residuals of the model do not respect the normality, independence and constant variance hypotheses. The distribution of residuals is asymmetric and leptokurtic and residuals present periods of strong volatility, alternating with periods of low volatility. Finally, Huber (1995) shows that the Wilkie model is biased using some data corresponding to the years of strong inflationary shocks in Great Britain (in 1920, 1940 and 1974). If these years are excluded from the regression, the coefficient of correlation between dividend yields and inflation is not significant anymore.

There are also theoretical criticisms again on the Wilkie model. Indeed, two central hypotheses of economic theory are not respected: the efficiency of financial markets and the absence of an arbitrage opportunity. However, it turns out that these two criticisms are challenged in a crisis period and that most asset management models do not respect these two principles.

To create a crisis generator adapted to the current context, several modifications have been brought by Hervé Fraysse to the basic Wilkie structure:

– Modification of used data:

To take into account the critique of Kitts and to obtain parameters reflecting the current economic context, monthly data over the period (2000, 2011) are considered. This period seems appropriate to take into account the changes of politico-economic context which we evoked earlier. In addition, having defined this period, the consideration of annual or quarterly data is not appropriate to obtain reliable modeling.

– Modification of the modeling of the inflation:

First, the core inflation is modeled, whereas the inflation was modeled by Wilkie because during the last 10 years, the prices of foodstuffs and the oil price have been very volatile.

Moreover, a study of temporal series considering all the processes (SARIMA) leads us to retain the following process SARIMA (1,1,1(0,0,1))12. The ACF and PACF give rise to a seasonality of period 12, validated by the plot of the Fourier spectrogram.

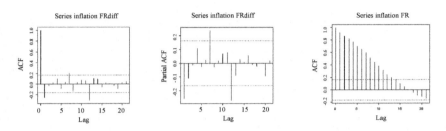

Figure 3.3. *Autocorrelation and partial autocorrelation plot of inflation*

From a macroeconomic point of view, it is clear that the inflation index is sensitive to the seasonality. Indeed, prices undergo different variation seasons and months: in December, for example, Christmas has a cyclic effect on prices.

– Modification of the residuals distribution:

A main problem is that the residuals obtained are usually not normal. Consequently, it is more precise to model the residuals differently. A solution to this problem consists of modeling the volatility of residuals using a GARCH process: Wilkie proposed this modeling in the 1995 version. Studies proved that the use of this method was less effective than the initial method. Hervé Fraysse proposed to model residuals by a hybrid Pareto distribution. This distribution is built in the following way: the center is modeled by a normal law and the tails of distribution are modeled by a generalized Pareto distribution. By fixing conditions of continuity, derivability and density, a statistical law depending on four parameters (average, variance and parameters of tail) is obtained (see Figure 3.3).

The modeling by a hybrid Pareto distribution is effective mainly in periods of crisis. Indeed, for these periods, the normal law leads to an underestimate of extreme values.

– Integration of jumps in the equation:

In the crisis generator created, a jump function is also added to the Wilkie structure. The function added is a temporary change function as shown in the graph below.

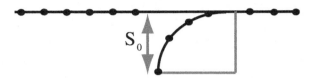

Several forms of intervention function have been studied in the ISUP paper to finally retain the transient change function.

This function allows the user to take into account different kinds of scenario in the projection. It is possible to define a more precise intervention function by considering a temporary change function with a level shift, but this seems more complex.

It is through three jump parameters (amplitude, probability of occurrence and parameter of reduction), integrated during the projection by an interface, that the fund manager can integrate his macroeconomic analysis. The values of the coefficients are determined "according to experts". For example, Figure 3.4 illustrates the evolution of inflation through different kinds of scenario.

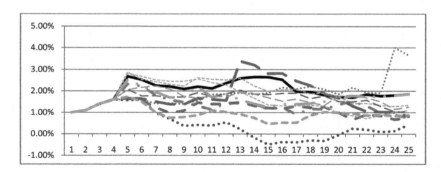

Figure 3.4. *Chosen function: Temporary change function. For a color version of the figure, see www.iste.co.uk/clement/forecasts.zip*

It is clear that the jump function is taken into account very well: indeed, we can see jumps in the projection.

3.5.4.5. *Multi-strategy diversification for equity management*

Core satellites can be based on ESG: the core portfolio can be managed using the most probable scenario; a satellite can be based on the second most probable scenario; its size compared to the core size can be in proportion to their respective probability; other satellites can be managed on other scenarios; for example, if the probability assigned to an oil price shock becomes significant, then the satellite can be managed to be protective in this case; an analogous protective satellite could be sized as a function of a food price big move with a major climate problem. Some satellites could be managed with a strategy designed to minimize one or several risk factors.

To diversify an asset management in such a way helps the Sharpe ratio performance, even in a big crisis; it is easier to work with core satellite management. Satellites can be designed to diversify as efficiently as possible; this diversification can be studied for several scenarios, the most probable. ESG can be used differently: probabilities assigned to crisis scenarios can be summed up to help the manager to decide the amount to be put in a low-risk satellite.

A core satellite can diversify with satellites managed with different processes quoted previously or a simpler process; for example, a satellite can be managed by diminishing small exposure as high-risk scenario probabilities are increasing. Small caps and other illiquid stocks are dangerous when a crisis is looming. A resilient satellite can be managed with very liquid stocks that stay liquid in a crisis; a minimum variance satellite can be made or it could be a minimum risk satellite based on another measure of risk. Risk parity can be made with a measure of risk decided by the manager; all measures of risk can be computed using the ESG; a number of strategies can help diversification and can be studied using risk measures calculated with an ESG. The manager has a large choice of methods; the ESG helps him/her to choose from among a variety of methods those that fit with the best diversification for the most probable scenarios as simulations are possible for these scenarios.

More generally, diversifying management methods gives seemingly better more resilient diversification than country or even industry diversification. The ESG can help the manager to allocate budgets and risk budgets to some management methods quoted previously or some variants: risk parity, rather low risk, can have variants with more or less active or conditional risk parity; the application of the modern portfolio theory (MPT) can require the use of one beta or two betas (traditional beta and/or smart beta).

Portfolios that are not managed as core satellites can have some specialized pockets. It is less convenient to subcontract a pocket than a satellite; therefore, a core satellite can have more satellites. Nevertheless, the manager can give more allocations to a pocket specialized for resilient stocks with good yield and good value that have always a good liquidity when crisis scenario probability is becoming important.

Risk valuation using a economic scenario generator does not rely on the past as risk factors valuations. Risk factors obtained by data analysis (such as principal components analysis and agglomerative hierarchical clustering) or by modeling (for example, based on multiple regression analysis) could be completely upset by any important crisis. Risk valuation based on scenarios will be changed because scenarios' probabilities will be changed but not upset. Any kind of crisis can be included; they cannot be taken into account by risk factor models that are used by many asset management companies; some use up to 50 risk factors which means many correlations; however, during the crisis, correlations shoot up above 80%; other companies use five risk factors with size as the main risk factor because liquidity risk is very much linked to size. Liquidity risk is very difficult to measure; therefore, a process using size could avoid this problematic measure.

3.5.4.6. *Conclusion about equity management*

All asset management processes can be improved by an ESG. Scenario parameters can be diversified by stochastic processes and various modelings (ARMA, ARIMA and GARCH). Scenarios can be tested and also aggregated in risk measures and risk premiums; they can be classified according to probability importance.

The Bayesian approach, based on scenarios, gives many ways to study behaviors and even irrational behaviors; but many, even most, are based on macroeconomic relationships that can be cyclical. All these scenarios' uses can be applied to bond markets and bond management, which will be studied in the next section.

Scenarios can be used for microeconomic studies to evaluate companies' risks that can be used for bond management and equity management, and for all kinds of assets a thorough financial analysis of companies with earnings forecasts on each scenario is a time-consuming process that gives better valuations.

3.6. Resilient bond portfolio building

As with skewness, which is an essential resiliency measure, convexity is an essential resiliency measure in the case of a sharp rise in bond rates; risk budgeting with the most important allocation to short-term bonds is better in a bear bond markets. Diversification of issuer quality could be done by risk budgeting and be helpful in many scenarios. Convexity with risk budgeting is convenient to allocate very short bonds and very long bonds in the right proportion to obtain a maximum convexity.

When, after more than 30 years of a bull bond market, will a bear market come with sharp rates rise? This will happen when the abnormality of rates near zero will no longer be needed to avoid deflation. It will be necessary to minimize the sensitivity to rate hikes with previously quoted methods; many managers will try to anticipate reducing strongly these risks at better prices. So, volatility could be high, strong convexity will be necessary.

In the USA, chair of the federal reverse Janet Yellen mentioned that this turning point could be possible in 2015; elsewhere, it is unforeseeable. To this turning point, there will be uncertainty about the kind of rate trend to be seen. Is it necessary to allocate to very short-term bonds a big portfolio proportion? Is it necessary to keep very long bonds in the right proportion to have a maximum convexity? If a quick and sharp turnaround was to be seen, is it better to hedge with options and some derivative products? If these products are already expensive, maximum convexity is more advisable.

To make that kind of choice, an economic scenario generator is helpful taking into account all kinds of rate turnarounds. In each scenario, the computation of the portfolio's current yield and expected return as well as risk measures implies a rational choice. The ability to transform portfolios quickly at the right time when so many managers will try to completely reverse their allocations is not easy to foresee; some examples can be found in the past: 1994 was the only important correction, a corrective short bear market; the rates market changed so much all along this 30 year bull market that the past is not easy to use to build scenarios. Transposition is needed.

An oil shock coming from an unforeseeable Middle East or a coincidence of a climate problem (with food price upsurge) and a war in the Middle East would be a worst-case scenario with the kind of inflationary process which implies a financial, economic, social and political crisis process. If a worst-case scenario for bond markets was to be seen because growth and/or inflation returned, fiercely, portfolio conditional value at risk is the best risk measure to judge that kind of risk which could last longer than the Vietnam War inflation. In this worst case, to understand which the resilient issuers are is not easy. Financial analysis of an issuer has to be done over the long term as with the resilience study of stock issuers. Which are the most liquid issues? Finding which kind of hedge is not too expensive over the medium term is essential. To be well prepared for that kind of scenario without being mesmerized by this worst case, nothing is better than an economic scenario generator with endeavors to determine a rational probability for such kinds of scenario.

3.6.1. *How could a rate trend turn around?*

Since 1981, world rate trends have been on a downtrend, so bond management methods were adapted to long duration portfolios to perform well in bull bond market scenarios optimizing the allocation of different kinds of bond; some reactions to these trends were seen in 1994; for example, some defensive methods were also studied well using convexity with the Barbell kind of portfolio optimizations using short-term bonds and long-term zero coupon bonds.

Now, with rates near zero and many bad geopolitical scenarios and some worst-case scenarios, it is necessary to study how to make a resilient bond portfolio in these cases and in some rate rebound cases (bigger than the 1994 rebound). To protect bond portfolios in these scenarios, caps (if anticipation is sufficient to buy these at a normal price) are very important tools because their value has an anti-correlation to bond prices that diminish the duration and increase the convexity of portfolios in which they appear. Bond management methods in the case of sharp rebound and in the case of markets taking account of an upward rate trend have not been studied very much as is normal after such a long bull bond market lasting more than 30 years.

How can we follow a rate trend turning around? It is easy to follow the primary market volume, issues that are difficult to sell and issue failures that could be the first step of risk aversion and decreasing liquidity. On the secondary market, volume statistics are often not available, but market makers' behaviors are very helpful to foresee problems; these behaviors are influenced by inventory funding costs and problems; capital constraints could be important indicators; bid and ask spreads are helpful in understanding the risk aversion of market makers and liquidity changes in each bond market segment. Statistical studies can be helpful (for example, autocorrelation).

The convertible bonds market can be a leading indicator as it was after the Lehman failure because it is a low liquidity market. Convertible arbitrage hedge funds endured big repayments, immediately after this failure; as they were often leveraged, they were obliged to sell at very low prices; this market became illiquid before the bonds market.

To survey liquidity of all bonds markets, a ratio is useful: (ask price – bid price)/(ask price + bid price). If quantities are available, to add "ask" quantity and "bid" quantity is useful; this quantitative approach is useful in any rate turnaround, but especially in the most dangerous ones. Volatility implied by various markets could help to foresee short-term important volatility changes. Fear propagation could be quick to change the volatility level and to imply lower liquidity in many markets.

3.6.2. *A protective bond management*

In an ISUP working paper by Hoeung Sakda [SAK 98], there is a study of caps (simulations on 2 and 5 years caps) qualities and drawbacks based on the formula by Briys, Crouhy and Schoebel (BCS formula) which is compared to Fisher Black's methodology. There is also in this working paper a study of cap replication by a special asset management method which is a very protective and resilient bond management method. Optimizations can be done on some or all scenarios generated by an ESG. To obtain cap premium and the total amount of the portfolio protected by cap replication, parameters of the BCS formula are needed: rate trends and volatilities with cap characteristics and transaction fees. Simulations of cap protected portfolios under many scenarios of rate increase give the possibility to optimize.

Caps built and sold by an intermediary are often difficult to sell at a reasonable price by the institution that holds, so, replication them which is a protection that can be managed with low transaction fees, is a very good protective policy. A cap is the sum of caplets which are put on zero coupon bonds. Caplets can be used to erase a great proportion of bank balance sheet mismatches. For insurance companies, to earn money with caps and caplets is convenient when an upward rate move implies bond value losses.

Fisher Black showed how to erase a bond portfolio risk locally at some point in time shorting a future and owning a call; the call price follows a process:

$$\frac{DF}{F} = a * dt + b * dz, with:$$

where:

   − F, a call price;

   − a, an expected return;

   − b, a standard deviation;

   − dz, a standard Brownian variable, if short-term rate and volatility are stable.

This is used to protect the raw material position, but it does not fit the cap protective effect in an interest rate rise very well. Nevertheless, comparison between this method and the BCS formula, which is more complex but fits the cap protective effect in an interest rate rise well, is interesting; many simulations were made with caps of 1–6 years with various move scenarios of the rate curve; this could be summarized as:

   − upward rate significant move gives a better gain if cap replication is done using the BCS formula compared to Fisher Black's method or a theoretical model; a swap could be used. Seesaw rate scenario simulations showed that the BCS cap replication is still better;

   − downward rate moves are less costly with a cap replication using Fisher Black's method.

There is also Leland's method taking account of transaction expenses and Hull and White's trinomial tree method and some Monte Carlo methods; this methodology will become popular and will expand when rates rise, and even with a rate rise trend, will be a reality which makes barbell convexity protection insufficient.

3.6.3. *ESG and bond management*

As English-speaking countries seem not very far from a period of significant upward move of rates, that kind of bond portfolio protective management could be important; an economic scenario generator is a help to work rationally on many scenarios of an uncertain rate trend turnaround with this protective management method.

The Wilkie ESG has an inflation rate process and real rate process; there is a model for the difference = long-term rate less short-term rate. This was fit to positive inflation, but seems also to be good for negative inflation rate. Zero coupon bond prices are based on a stochastic actualization factor in the Brennan and Xia ESG. In the Ahlgrim ESG, inflation rate is an Ornstein–Uhlenbeck process as in Vasicek's one-factor model; zero coupon bond price is evaluated with Hull and White's two-factor formulas; inflation rate and real interest rate are supposed to be independent, i.e. valuated separately. All ESGs have qualities and drawbacks on bond markets. For worst-case scenarios with oil shocks, the Wilkie ESG seems well fit with inflation shock coming from oil as a stable Poisson process. It is possible to use extrema theory.

Computation of many risk measures can be done successively with various degrees of protection by caps, and other protective methods to choose the best fit to the portfolio management aims and constraints. Some interesting papers were based on a simple autoregressive scenario generator built by Wilkie:

– a working paper (see the working papers listed in the Bibliography) in 2010–2011 added two Poisson processes, a food price shock and an oil price shock without a mean reversing process. The created generator is adapted to strong shocks, unforeseeable "black swan" shock, occurring seldomly to exceptionally moving rate curves by a very quick translation with a comeback (or a not so quick translation come back). Vasicek's model has a mean reversing term or two;

– a dissertation [FRA 12] modified the inflation modeling of Wilkie (1985) and equation residue distribution using Pareto's law. The created generator is adapted to a strong shock process, not as unforeseeable as "black swans". Two mean reversing terms are useful. Hull and White's model could be good to take into account a mean reversing process of long bonds and rather short bonds;

– another working paper (see Bibliography) introduced auto regressions with jumps and used copulas to link main variables. The created generator is adapted for moderate and rather frequent shocks process which modifies rate curve rather frequently; Heath, Jarrow and Morton's complex model is well fitted in this case.

To optimize any optional and/or derivative protection for turbulent periods, a simple Vasicek's model is well adapted with a few parameters which avoid parameter instability, especially in the case of "black swan" extreme shock. Even in less turbulent periods, but with frequent significant jumps – "black butterflies" shocks – it is better not to use too many parameters. Not so turbulent periods give the ability to model with more numerous parameters; so, a more precise description of bond markets is possible and useful if parameters are not too unstable. In this book, worst-case scenarios are a main topic; this means that turbulent periods are a very important topic; with a scenario generator, a good bond model is necessary to build a good protection, to measure it in order to diminish risks.

When an increased probability is put on oil price shock or some other extreme scenarios, it is necessary to increase portfolio resilience that can be obtained with an increased proportion of first quality issuer rather than short bonds of solid countries' treasuries of non-cyclical great resilient firms. An increased flexibility can be obtained with very liquid bonds which stayed liquid in the last crisis; the idea that bond illiquidity for many segments of rate markets could be back on an extreme scenario is being put into effect, even for not so extreme scenarios.

Convexity well monitored with a scenario generator is interesting to obtain a good return compared to competition, even in the case of a sharp move which can happen late in many scenarios; good monitoring implies risk budgeting to allocate risk with precision on very long bonds that will probably be illiquid in the next crisis and short ones chosen as mentioned previously to stay liquid at any time. Risk budgeting with CVaR constraint seems the best; some backtesting on the 1994 sharp bond rates rise would be helpful to define the risk budgets.

In a period of high indebtedness of many countries where public debt is the same size as GNP or higher, the state bill book matters; the amount of public debt coming to maturity each year matters. Budget deficit matters, primary deficit matters; if the receipts less expenditures balance including debt interest payment is negative, debt will go on increasing. Screening on all these figures is a process to list resilient sovereign issuers.

When endeavors to diminish the budget deficit, trade balance deficit, are lowering the growth rate of the economy, tax receipts go down; when rate paid on debt increases and adds to expenditures, the budget deficit is moving up. Some scenarios of a refinancing crisis can be modeled. When foreign investors shun state issues, national investors can be very worried and shun these issues as well. This troubled state asks for help from the European authorities and the IMF. To hold bonds on this issuer could be very costly.

If there is no lender of last resort willing or able to lend sufficiently, the troubled state can compel bond holders to conversion into perpetual bonds with or without haircuts; depositors can be compelled to buy state bonds; deposits can be frozen and various forms of compulsory subscriptions have been seen in Latin America. The optimum allocation will move toward flexibility to try to avoid haircuts and compulsory subscriptions; when that kind of Latin American crisis scenario has an increasing probability, allocation must be changed quickly to corporate bonds of the best quality issuer, moreover, issues of the best international issuer. If there is no way left out, at a time when probability of that kind of scenario is increasing, the only optimization that could be made by risk budgeting is screening as a function of issuer quality, liquidity record and bond issue ownership.

Illiquidity in monetary and bond markets came with the fear of systemic crisis in 2008; the good running of markets is based on trust: when trust disappears, markets fade and volume disappears. At the end of 2008, highly leveraged hedge funds willing to sell large amounts of bonds and/or convertible bonds were not able to sell:

– institutions were risk adverse, unwilling to buy;

– market makers were risk adverse, unwilling to trade bonds.

These operators still have a vivid remembrance of the psychological effect of closed markets, which could come back easily as interest rates are so low that they could rise a lot at once; some leading indicators of this move would imply a huge bear bond market and illiquidity and crash. So, many institutions would like to sell. Risk adverse traders will stop trading, remembering 2008. So, bond market liquidity could be a problem, even with a smaller crisis. For this kind of scenario, it is good to increase convexity with very long solid state bonds diversified by risk parity budgeting; for short bonds, risk parity could also be useful. Backtesting studies of the 1994 and the 1973–1980 inflationary period with utility function maximization showed that institutions would be better to include skewness and CVaR constraints, but CVaR and its qualities are not well-known concepts at the moment; if CVaR at that time of crisis is not better known, VaR could be used to replace CVaR.

The easiest arbitrage between two similar bonds for price reasons is obviously very usual and can be done without risk computation if the duration is similar. To increase skewness increases dissymmetry with more sensitivity to a price up move than the reverse; so, it increases convexity. Another skewness advantage appears when a list of portfolio sales to be done is established: co-skewness of a line to be sold gives a precise idea of the effect of the sale of this line on the utility function; co-skewness of a possible purchase gives the potential effect on the utility function; optimization under the transaction cost budget limit can be done as was described for an equities portfolio. A step-by-step screening is also possible. Precise skewness and co-skewness can be done with an economic scenario generator on a day-to-day basis, which is a big help for management.

3.7. Application

As for multi-strategy diversification for equity management, core satellite management is the easier methodology to subcontract and diversify to various asset classes: equities, bonds, derivative products, structured products and other tradable assets. With precise risk valuations for many kinds of risk measure, an ESG is well adapted to help decide allocation.

Figure 3.5 summarizes the process to be settled for a resilient asset manager.

3.8. Conclusion

As a summary of the optimization methodology, there is static optimization and dynamic optimization:

– static optimization for passive management or for benchmarked portfolios implies tracking error minimization as risk criteria;

– static optimization on other kinds of stock management can be done on risk/return criteria, on utility function maximization with risk measures chosen among moments and/or extreme risk measures (VaR, CVaR and TVaR);

– constraints can be set;

– dynamic optimization can be done with dynamic programming or martingale methodology.

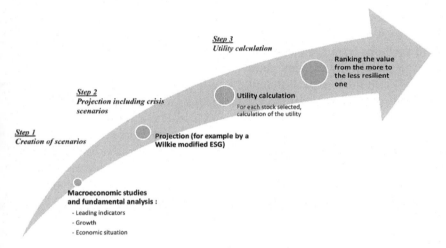

Figure 3.5. *Evolution of inflation through different kinds of scenario*

Asset managers have to make a choice from amongst many allocation processes and many risk processes; they need insightful comparisons. Risks change so frequently in size and nature that it is difficult to decide which process is the best. Backtestings as those done in the ENSAE working papers give comparisons, but, on the whole, these comparisons are elaborate and sometimes intricate. Nevertheless, there is a clear and simple conclusion: to estimate precisely risks of many assets, with many scenarios possible, nothing is better than an economic and financial scenario generator. Changing from a generator for an ordinary cyclical period to a crisis scenario generator implies difficulties to manage transaction costs resulting from a switch of portfolio assets. This large and costly turnover has to be done only if the manager has the firm belief that a great crisis (a black swan) is coming; however, a crisis process that could mean many crises (black butterflies) is not easy to forecast. Thorough experience is needed and many scenarios are to be studied to choose a process; a good ability to forecast and judge a beginning crisis is needed to switch from an ordinary ESG to a crisis ESG with good results. The use of a regime-switching process as described in section 3.4.1 is not very costly if the crisis is an ordinary one or even if crisis does not come. Risk parity with resilient stocks could still be less costly in the case of a false crisis warning.

As the number of risks increases, thinking more precisely with future crisis scenarios, with experts say on geopolitical risk scenarios, is necessary; many management processes have been quoted, and some were more precisely studied; choices made by the managers take into account short-term and medium-term scenario; the long term is too fuzzy to be defined by scenarios; a philosophy to think about the long term is a better approach. In a quiet period, a rational expectation hypothesis is a useful approximation, but, imperfect information, imperfect knowledge and imperfect rationality involve many scenario studies in turbulent periods. Learning in turbulent periods is an uncertain process that could imply stampedes to avoid some perceived crisis. Sharp turns in risk aversion could amplify instability and crises. Reinhart and Rogoff [ROG 09] showed that common features can be found among crises, but it is an *ex post* study that can show these features that are considered among scenarios, but not forecasted.

When a crisis is perceived, it is not possible to judge the kind and the pattern of a crisis because many scenarios are possible. It is not possible to manage with too many scenarios in mind; but to face a dangerous period, core satellite management can take account precisely of some scenarios as mentioned previously; to manage with resilient pockets inside a portfolio is also possible to diversify at a time when correlation is bluntly up; core satellite with some subcontracted satellite devoted to some risk covering can be studied as improved asset liability equilibrium. Wars erupting in Asia, the Middle East or Eastern Europe are bigger risks than cyclones or earthquakes; the use of a Poisson process and Pareto's law for casualties is classic; using them for wars and other geopolitical casualties that cannot be forecasted is less classic. The Bayesian approach with probability-assigned scenarios is a clear and simple process to calculate any kind of risk measures needed for any kind of management process.

Conclusion

As a general conclusion of many backtest simulations, a process based on the methods previously quoted must be settled with the choice of a method previously quoted. Having scrutinized many scenario categories and many crisis scenarios, it is clear that when a crisis seems to be looming it is necessary to frequently update scenario probabilities and to update the scenario list. Markov regime switching is well adapted to this kind of situation when a crisis risk increases, but this does not mean that this risk cannot disappear. To switch back if a crisis scenario probability is clearly decreasing can be done rationally. An easier method to manage a hazardous crisis risk is to create a resilient asset management pocket devoted to resilient stocks; this pocket can easily be decreased or increased as a function of the crisis risk. After the study of resiliency, most resilient stocks need to be listed in various crisis scenarios; from amongst this list, the asset manager decides the number of stocks to be bought; the same sum is invested in each stock. When many crisis scenarios can be seen, it seems that the simplest method is the best, but there is a diversification problem at a time of crisis when all correlation coefficients are increasing. This result appeared with some studies on the resilience of asset management to a crisis. It is very helpful to monitor skewness and several other dissymmetry measures in these comparisons.

In a more mature stage of the crisis process, when the systemic crisis risk seems sizable, a change of paradigm can be decided by the asset manager to take into account many crisis scenarios in risk valuation, including or not systemic crisis scenarios; it is the time when quants have to be quick to change their models to take into account the quick change of risk measures and market priorities; strong risk aversion linked to fear and the resulting illiquidity are very difficult to model during a crisis. At this stage, *quant* behaviors could be very different from average behaviors; it is easier to give subjective probabilities to scenarios. As the crisis process unfolds, the number of scenarios with very large probabilities diminishes quickly. When less than a dozen scenario probabilities sum up to 75%, it is an interesting time for the core satellite management to find a subcontractor well fit to each scenario; this kind of management could be very convenient on the way out of the crisis to play, in various satellites, the rebounds of different categories of stocks (depressed value stocks, cyclical industrial, small stocks, etc.) of various countries and different industry sectors.

There are worst-case risks that could imply many scenarios because the European Central Bank and the International Monetary Fund are not lenders of last resort. Indeed, the IMF is an institution that was created at Bretton Woods that has to be increased in size and capability; but, regardless, the International Monetary Fund and some other financial institutions would try to help as central banks; each government and each treasury would try to avoid systemic crisis; coordination is a low-probability scenario. Many reactions in Germany, and in other countries, could be disruptive. To cover these kinds of multi-step processes methodically, some "trees" with a bad or a good piece of news at each node are helpful; it is also interesting to study a two-player game (northern and southern Europe or Germany leaving the Euro-area alone). Multi-player games with imperfect information and some cooperation seem too intricate. Geopolitical and climatic worst-case scenarios could also be difficult to study.

Former International Monetary Fund general manager Michel Camdessus has already quoted reform opportunities in crisis, worried, in recent conferences, that globalization without a stronger G20 capability to induce regulation improvements could imply other systemic crises. For asset managers this means, also, an opportunity to outperform the market index study of past crisis scenarios and possible future crisis scenarios linked to the trend of increased complexities of the connected financial world. This lengthy process has to be carried out in an ordinary period when rationality is not impaired by fear. Capitalism is a system of management by crisis; this definition is especially important since the end of the Bretton Woods monetary system; to understand crisis possibilities and crisis processes with the help of a crisis scenario generator is a method to avoid panicky behaviors, a method to outperform competitors.

Appendix

A.1. Appendix

A.1.1. *Inflation rates*

World

Country Nb	County name	Number	2016	2017	2018	2019	2020	End of cycle	Long term tendency	Indicator
1	WORLD	111	3,2	3,0	3,2	3,3	3,3	Slight economic downturn in 2016	Optimistic	↑
1	WORLD	112	3,1	2,8	2,9	3,0	3,2	Sharp economic downturn in 2016	Optimistic	↑
1	WORLD	113	3,0	2,0	2,5	2,7	2,9	Risk of recession in 2016	Optimistic	↑
1	WORLD	121	3,0	2,9	3,0	3,2	3,3	Slight economic downturn in 2017	Neutral	=
1	WORLD	122	3,0	2,8	2,5	2,8	3,1	Sharp economic downturn in 2017	Neutral	=
1	WORLD	123	3,0	2,7	1,8	2,0	2,6	Risk of recession in 2017	Neutral	=
1	WORLD	131	2,0	1,8	1,6	1,0	1,2	Slight economic downturn in 2018	Stagnation	→
1	WORLD	132	2,0	1,8	1,4	0,5	0,7	Sharp economic downturn in 2018	Stagnation	→
1	WORLD	133	2,0	1,7	1,0	-1,0	-2,0	Risk of recession in 2018	Stagnation	→
1	WORLD	141	1,5	1,4	1,3	1,1	1,0	Slight economic downturn in 2019	Deflation	↓
1	WORLD	142	1,5	1,4	1,2	0,5	0,0	Sharp economic downturn in 2019	Deflation	↓
1	WORLD	143	1,0	-0,5	-1,0	-2,0	-3,0	Risk of recession in 2019	Deflation	↓

Europe

Country Nb	County name	Number	2016	2017	2018	2019	2020	End of cycle	Long term tendency	Indicator
2	GERMANY	211	1,2	1,3	1,2	1,2	1,2	Slight economic downturn in 2016	Optimistic	↑
2	GERMANY	212	0,0	0,0	0,2	0,5	0,7	Sharp economic downturn in 2016	Optimistic	↑
2	GERMANY	213	-0,5	-1,0	-0,5	0,0	0,6	Risk of recession in 2016	Optimistic	↑
2	GERMANY	221	0,8	0,6	0,4	0,6	0,9	Slight economic downturn in 2017	Neutral	=
2	GERMANY	222	0,2	0,0	-0,2	0,0	0,5	Sharp economic downturn in 2017	Neutral	=
2	GERMANY	223	-0,5	-1,8	-0,8	0,0	0,4	Risk of recession in 2017	Neutral	=
2	GERMANY	231	0,7	0,6	0,3	0,0	0,2	Slight economic downturn in 2018	Stagnation	→
2	GERMANY	232	0,1	0,0	-0,3	0,2	0,4	Sharp economic downturn in 2018	Stagnation	→
2	GERMANY	233	-0,6	-2,0	-1,0	0,0	1,0	Risk of recession in 2018	Stagnation	→
2	GERMANY	241	0,5	0,4	0,3	0,2	-0,2	Slight economic downturn in 2019	Deflation	↓
2	GERMANY	242	0,4	0,3	0,1	0,0	-1,0	Sharp economic downturn in 2019	Deflation	↓
2	GERMANY	243	-0,7	-2,5	-1,5	-0,3	0,4	Risk of recession in 2019	Deflation	↓
3	FRANCE	111	1,0	1,1	1,0	1,0	1,2	Slight economic downturn in 2016	Optimistic	↑
3	FRANCE	112	0,0	0,0	0,2	0,5	0,7	Sharp economic downturn in 2016	Optimistic	↑
3	FRANCE	113	-0,5	-1,0	-0,5	0,0	0,6	Risk of recession in 2016	Optimistic	↑
3	FRANCE	121	0,8	0,6	0,4	0,6	0,9	Slight economic downturn in 2017	Neutral	=
3	FRANCE	122	0,2	0,0	-0,2	0,0	0,5	Sharp economic downturn in 2017	Neutral	=
3	FRANCE	123	-0,5	-1,8	-0,8	0,0	0,4	Risk of recession in 2017	Neutral	=
3	FRANCE	131	0,7	0,6	0,3	0,0	0,2	Slight economic downturn in 2018	Stagnation	→
3	FRANCE	132	0,1	0,0	-0,3	0,2	0,4	Sharp economic downturn in 2018	Stagnation	→
3	FRANCE	133	-0,6	-2,0	-1,0	0,0	1,0	Risk of recession in 2018	Stagnation	→
3	FRANCE	141	0,5	0,4	0,3	0,2	-0,2	Slight economic downturn in 2019	Deflation	↓
3	FRANCE	142	0,4	0,3	0,1	0,0	-1,0	Sharp economic downturn in 2019	Deflation	↓
3	FRANCE	143	-0,7	-2,5	-1,5	-0,3	0,4	Risk of recession in 2019	Deflation	↓
4	ITALY	111	1,0	1,1	1,0	1,0	1,2	Slight economic downturn in 2016	Optimistic	↑
4	ITALY	112	0,0	0,0	0,2	0,5	0,7	Sharp economic downturn in 2016	Optimistic	↑
4	ITALY	113	-0,5	-1,0	-0,5	0,0	0,6	Risk of recession in 2016	Optimistic	↑
4	ITALY	121	0,8	0,6	0,4	0,6	0,9	Slight economic downturn in 2017	Neutral	=
4	ITALY	122	0,2	0,0	-0,2	0,0	0,5	Sharp economic downturn in 2017	Neutral	=
4	ITALY	123	-0,5	-1,8	-0,8	0,0	0,4	Risk of recession in 2017	Neutral	=
4	ITALY	131	0,7	0,6	0,3	0,0	0,2	Slight economic downturn in 2018	Stagnation	→
4	ITALY	132	0,1	0,0	-0,3	0,2	0,4	Sharp economic downturn in 2018	Stagnation	→
4	ITALY	133	-0,6	-2,0	-1,0	0,0	1,0	Risk of recession in 2018	Stagnation	→
4	ITALY	141	0,5	0,4	0,3	0,2	-0,2	Slight economic downturn in 2019	Deflation	↓
4	ITALY	142	0,4	0,3	0,1	0,0	-1,0	Sharp economic downturn in 2019	Deflation	↓
4	ITALY	143	-0,7	-2,5	-1,5	-0,3	0,4	Risk of recession in 2019	Deflation	↓
5	SPAIN	111	0,8	0,7	0,9	1,5	2,0	Slight economic downturn in 2016	Optimistic	↑
5	SPAIN	112	0,7	0,5	0,6	0,7	0,8	Sharp economic downturn in 2016	Optimistic	↑
5	SPAIN	113	0,6	-0,2	0,0	0,5	0,7	Risk of recession in 2016	Optimistic	↑
5	SPAIN	121	0,5	0,4	0,3	0,7	0,8	Slight economic downturn in 2017	Neutral	=
5	SPAIN	122	0,5	0,4	0,0	0,5	0,7	Sharp economic downturn in 2017	Neutral	=
5	SPAIN	123	0,5	0,0	-0,5	-0,2	0,5	Risk of recession in 2017	Neutral	=
5	SPAIN	131	0,3	0,3	0,2	0,4	0,8	Slight economic downturn in 2018	Stagnation	→
5	SPAIN	132	0,3	0,3	0,0	0,5	0,7	Sharp economic downturn in 2018	Stagnation	→
5	SPAIN	133	0,3	0,3	-0,5	-0,6	0,3	Risk of recession in 2018	Stagnation	→
5	SPAIN	141	0,1	0,2	0,2	0,0	0,2	Slight economic downturn in 2019	Deflation	↓
5	SPAIN	142	0,1	0,1	0,1	-0,4	-0,5	Sharp economic downturn in 2019	Deflation	↓
5	SPAIN	143	0,1	0,1	0,0	-2,0	-2,5	Risk of recession in 2019	Deflation	↓
6	GREECE	111	2,4	2,4	2,5	2,6	2,8	Slight economic downturn in 2016	Optimistic	↑
6	GREECE	112	2,0	2,0	1,8	2,2	2,3	Sharp economic downturn in 2016	Optimistic	↑
6	GREECE	113	-0,3	-0,5	-0,2	0,0	0,3	Risk of recession in 2016	Optimistic	↑
6	GREECE	121	2,3	2,2	2,0	2,2	2,3	Slight economic downturn in 2017	Neutral	=
6	GREECE	122	2,3	2,0	1,5	2,0	2,5	Sharp economic downturn in 2017	Neutral	=
6	GREECE	123	2,3	1,0	-0,5	-0,3	0,0	Risk of recession in 2017	Neutral	=
6	GREECE	131	2,0	2,2	2,1	1,9	2,0	Slight economic downturn in 2018	Stagnation	→
6	GREECE	132	2,0	2,0	1,8	1,0	1,0	Sharp economic downturn in 2018	Stagnation	→
6	GREECE	133	2,0	1,9	-0,8	-1,5	-0,5	Risk of recession in 2018	Stagnation	→
6	GREECE	141	0,9	1,0	0,9	0,7	1,0	Slight economic downturn in 2019	Deflation	↓
6	GREECE	142	0,9	0,8	1,0	0,0	0,5	Sharp economic downturn in 2019	Deflation	↓
6	GREECE	143	0,9	0,8	0,4	-2,0	-1,4	Risk of recession in 2019	Deflation	↓

English-speaking countries

Country Nb	County name	Number	2016	2017	2018	2019	2020	End of cycle	Long term tendency	Indicator
7	UK	111	2,3	2,1	2,3	2,5	2,7	Slight economic downturn in 2016	Optimistic	↑
7	UK	112	1,8	1,5	1,8	2,3	2,5	Sharp economic downturn in 2016	Optimistic	↑
7	UK	113	1,2	1,0	1,5	1,7	2,0	Risk of recession in 2016	Optimistic	↑
7	UK	121	2,5	2,4	2,1	2,3	2,5	Slight economic downturn in 2017	Neutral	=
7	UK	122	2,5	2,4	1,5	1,8	2,0	Sharp economic downturn in 2017	Neutral	=
7	UK	123	2,5	1,0	1,4	1,6	1,7	Risk of recession in 2017	Neutral	=
7	UK	131	2,2	2,1	2,0	2,2	2,0	Slight economic downturn in 2018	Stagnation	→
7	UK	132	2,2	2,0	1,6	1,3	1,6	Sharp economic downturn in 2018	Stagnation	→
7	UK	133	2,2	1,2	0,0	0,2	0,3	Risk of recession in 2018	Stagnation	→
7	UK	141	2,0	2,2	2,0	1,8	2,2	Slight economic downturn in 2019	Deflation	↓
7	UK	142	2,0	2,0	1,8	1,0	1,2	Sharp economic downturn in 2019	Deflation	↓
7	UK	143	2,0	1,8	0,9	-0,4	0,0	Risk of recession in 2019	Deflation	↓
8	USA	111	2,3	2,3	2,5	2,6	2,7	Slight economic downturn in 2016	Optimistic	↑
8	USA	112	1,9	1,4	1,6	2,0	2,5	Sharp economic downturn in 2016	Optimistic	↑
8	USA	113	1,8	1,0	1,4	1,9	2,4	Risk of recession in 2016	Optimistic	↑
8	USA	121	2,6	2,4	2,5	2,6	2,6	Slight economic downturn in 2017	Neutral	=
8	USA	122	2,6	2,0	2,3	2,4	2,5	Sharp economic downturn in 2017	Neutral	=
8	USA	123	2,6	2,4	1,0	1,2	1,9	Risk of recession in 2017	Neutral	=
8	USA	131	2,0	2,0	2,0	2,0	2,5	Slight economic downturn in 2018	Stagnation	→
8	USA	132	2,0	1,9	1,6	1,8	2,3	Sharp economic downturn in 2018	Stagnation	→
8	USA	133	2,0	2,2	2,0	1,3	1,0	Risk of recession in 2018	Stagnation	→
8	USA	141	1,7	1,7	1,8	1,6	1,7	Slight economic downturn in 2019	Deflation	↓
8	USA	142	1,7	1,8	1,8	1,4	1,5	Sharp economic downturn in 2019	Deflation	↓
8	USA	143	1,7	1,9	2,1	1,3	1,0	Risk of recession in 2019	Deflation	↓

India

Country Nb	County name	Number	2016	2017	2018	2019	2020	End of cycle	Long term tendency	Indicator
13	INDIA	111	5,5	5,4	6,0	7,0	8,0	Slight economic downturn in 2016	Optimistic	↑
13	INDIA	112	5,0	4,5	5,5	6,6	7,8	Sharp economic downturn in 2016	Optimistic	↑
13	INDIA	113	2,0	1,5	3,0	6,0	7,0	Risk of recession in 2016	Optimistic	↑
13	INDIA	121	6,4	6,2	6,0	7,0	7,5	Slight economic downturn in 2017	Neutral	=
13	INDIA	122	6,4	5,4	5,0	5,5	6,5	Sharp economic downturn in 2017	Neutral	=
13	INDIA	123	6,4	5,0	3,0	4,0	5,0	Risk of recession in 2017	Neutral	=
13	INDIA	131	6,0	5,7	6,0	7,0	6,0	Slight economic downturn in 2018	Stagnation	→
13	INDIA	132	6,0	4,0	2,0	3,0	5,0	Sharp economic downturn in 2018	Stagnation	→
13	INDIA	133	6,0	3,0	0,0	-0,2	1,0	Risk of recession in 2018	Stagnation	→
13	INDIA	141	4,0	5,0	4,0	3,0	4,0	Slight economic downturn in 2019	Deflation	↓
13	INDIA	142	4,0	3,0	2,5	2,0	3,0	Sharp economic downturn in 2019	Deflation	↓
13	INDIA	143	4,0	3,0	2,0	-0,5	-1,5	Risk of recession in 2019	Deflation	↓

Russia/Poland

Country Nb	County name	Number	2016	2017	2018	2019	2020	End of cycle	Long term tendency	Indicator
14	RUSSIA	111	5,0	5,0	5,0	5,0	4,5	Slight economic downturn in 2016	Optimistic	↑
14	RUSSIA	112	4,0	3,5	4,0	4,5	4,5	Sharp economic downturn in 2016	Optimistic	↑
14	RUSSIA	113	2,0	3,0	4,0	4,8	4,8	Risk of recession in 2016	Optimistic	↑
14	RUSSIA	121	4,3	4,2	4,3	5,0	5,5	Slight economic downturn in 2017	Neutral	=
14	RUSSIA	122	4,3	3,6	4,0	4,5	4,5	Sharp economic downturn in 2017	Neutral	=
14	RUSSIA	123	4,3	4,0	3,0	3,3	3,5	Risk of recession in 2017	Neutral	=
14	RUSSIA	131	3,2	3,5	3,3	3,0	3,7	Slight economic downturn in 2018	Stagnation	→
14	RUSSIA	132	3,2	3,3	2,8	3,0	3,5	Sharp economic downturn in 2018	Stagnation	→
14	RUSSIA	133	3,2	3,7	2,1	1,7	3,0	Risk of recession in 2018	Stagnation	→
14	RUSSIA	141	2,7	3,0	3,0	2,5	2,5	Slight economic downturn in 2019	Deflation	↓
14	RUSSIA	142	2,7	2,7	2,5	2,0	2,0	Sharp economic downturn in 2019	Deflation	↓
14	RUSSIA	143	2,7	2,5	2,0	1,0	0,0	Risk of recession in 2019	Deflation	↓
15	POLAND	111	1,8	1,5	1,8	2,2	2,5	Slight economic downturn in 2016	Optimistic	↑
15	POLAND	112	1,5	0,8	1,2	1,5	1,8	Sharp economic downturn in 2016	Optimistic	↑
15	POLAND	113	0,7	0,0	1,0	2,0	2,5	Risk of recession in 2016	Optimistic	↑
15	POLAND	121	2,5	2,4	2,3	2,5	2,5	Slight economic downturn in 2017	Neutral	=
15	POLAND	122	2,5	1,7	2,2	2,4	2,6	Sharp economic downturn in 2017	Neutral	=
15	POLAND	123	2,5	1,0	1,5	2,6	2,7	Risk of recession in 2017	Neutral	=
15	POLAND	131	2,0	2,1	1,8	1,7	2,2	Slight economic downturn in 2018	Stagnation	→
15	POLAND	132	2,0	2,2	1,3	1,8	2,1	Sharp economic downturn in 2018	Stagnation	→
15	POLAND	133	2,0	2,0	1,0	0,0	0,5	Risk of recession in 2018	Stagnation	→
15	POLAND	141	1,0	1,4	1,5	1,2	1,2	Slight economic downturn in 2019	Deflation	↓
15	POLAND	142	1,0	1,2	1,3	0,9	0,4	Sharp economic downturn in 2019	Deflation	↓
15	POLAND	143	1,0	0,7	0,4	0,0	-0,3	Risk of recession in 2019	Deflation	↓

Asia

Country Nb	County name	Number	2016	2017	2018	2019	2020	End of cycle	Long term tendency	Indicator
9	CHINA	111	2,5	2,3	2,5	2,7	3,0	Slight economic downturn in 2016	Optimistic	↑
9	CHINA	112	2,0	1,0	1,5	2,0	3,0	Sharp economic downturn in 2016	Optimistic	↑
9	CHINA	113	1,0	0,0	1,0	2,0	3,5	Risk of recession in 2016	Optimistic	↑
9	CHINA	121	2,8	2,7	2,5	2,8	3,0	Slight economic downturn in 2017	Neutral	=
9	CHINA	122	2,8	2,8	1,6	2,0	2,5	Sharp economic downturn in 2017	Neutral	=
9	CHINA	123	2,8	3,0	1,3	2,0	2,7	Risk of recession in 2017	Neutral	=
9	CHINA	131	2,2	2,1	1,9	2,0	2,3	Slight economic downturn in 2018	Stagnation	→
9	CHINA	132	2,2	2,1	1,4	1,8	2,2	Sharp economic downturn in 2018	Stagnation	→
9	CHINA	133	2,2	2,0	1,0	1,5	2,3	Risk of recession in 2018	Stagnation	→
9	CHINA	141	1,0	0,9	0,7	0,8	1,0	Slight economic downturn in 2019	Deflation	↓
9	CHINA	142	1,0	1,2	1,0	0,4	0,8	Sharp economic downturn in 2019	Deflation	↓
9	CHINA	143	1,0	1,5	2,0	0,0	1,0	Risk of recession in 2019	Deflation	↓
10	JAPAN	111	1,3	1,2	1,8	2,0	2,5	Slight economic downturn in 2016	Optimistic	↑
10	JAPAN	112	1,0	0,7	1,0	2,2	2,7	Sharp economic downturn in 2016	Optimistic	↑
10	JAPAN	113	0,5	1,0	0,6	1,0	2,0	Risk of recession in 2016	Optimistic	↑
10	JAPAN	121	1,9	1,9	1,7	2,4	2,6	Slight economic downturn in 2017	Neutral	=
10	JAPAN	122	1,9	1,8	1,5	1,8	2,4	Sharp economic downturn in 2017	Neutral	=
10	JAPAN	123	1,9	1,6	0,0	-0,5	1,0	Risk of recession in 2017	Neutral	=
10	JAPAN	131	1,6	1,6	1,4	1,3	1,6	Slight economic downturn in 2018	Stagnation	→
10	JAPAN	132	1,6	1,5	1,1	0,5	1,0	Sharp economic downturn in 2018	Stagnation	→
10	JAPAN	133	1,6	1,0	0,0	-1,0	0,0	Risk of recession in 2018	Stagnation	→
10	JAPAN	141	1,2	1,0	1,2	1,4	1,8	Slight economic downturn in 2019	Deflation	↓
10	JAPAN	142	1,0	0,5	0,8	1,0	1,5	Sharp economic downturn in 2019	Deflation	↓
10	JAPAN	143	0,2	0,0	-0,2	-1,5	-0,5	Risk of recession in 2019	Deflation	↓
11	SOUTHKOREA	111	2,0	1,8	2,0	2,2	2,4	Slight economic downturn in 2016	Optimistic	↑
11	SOUTH KOREA	112	1,7	1,3	1,7	2,2	2,4	Sharp economic downturn in 2016	Optimistic	↑
11	SOUTH KOREA	113	1,0	0,5	1,0	1,5	2,0	Risk of recession in 2016	Optimistic	↑
11	SOUTH KOREA	121	2,2	2,0	2,3	2,5	2,7	Slight economic downturn in 2017	Neutral	=
11	SOUTH KOREA	122	2,2	1,5	2,0	2,4	2,8	Sharp economic downturn in 2017	Neutral	=
11	SOUTH KOREA	123	2,2	1,9	0,4	0,8	1,2	Risk of recession in 2017	Neutral	=
11	SOUTH KOREA	131	1,9	1,9	1,7	1,9	2,0	Slight economic downturn in 2018	Stagnation	→
11	SOUTH KOREA	132	1,9	1,7	1,0	0,7	1,0	Sharp economic downturn in 2018	Stagnation	→
11	SOUTH KOREA	133	1,9	1,6	0,8	0,0	0,3	Risk of recession in 2018	Stagnation	→
11	SOUTH KOREA	141	1,2	1,4	1,6	1,4	1,4	Slight economic downturn in 2019	Deflation	↓
11	SOUTH KOREA	142	1,2	1,1	1,1	0,3	0,0	Sharp economic downturn in 2019	Deflation	↓
11	SOUTH KOREA	143	1,2	1,1	1,0	-0,3	-0,6	Risk of recession in 2019	Deflation	↓
12	AUSTRALIA	111	2,7	2,6	2,8	2,9	3,0	Slight economic downturn in 2016	Optimistic	↑
12	AUSTRALIA	112	2,5	2,2	2,4	2,6	2,9	Sharp economic downturn in 2016	Optimistic	↑
12	AUSTRALIA	113	2,2	1,0	1,5	1,7	2,0	Risk of recession in 2016	Optimistic	↑
12	AUSTRALIA	121	2,8	2,7	2,5	2,7	2,8	Slight economic downturn in 2017	Neutral	=
12	AUSTRALIA	122	2,8	3,0	2,8	3,0	2,8	Sharp economic downturn in 2017	Neutral	=
12	AUSTRALIA	123	2,8	2,6	2,0	2,6	3,0	Risk of recession in 2017	Neutral	=
12	AUSTRALIA	131	2,6	2,9	2,6	3,0	3,2	Slight economic downturn in 2018	Stagnation	→
12	AUSTRALIA	132	2,6	2,7	1,5	1,0	1,5	Sharp economic downturn in 2018	Stagnation	→
12	AUSTRALIA	133	2,6	1,0	0,5	0,2	0,0	Risk of recession in 2018	Stagnation	→
12	AUSTRALIA	141	2,4	2,4	2,2	2,0	1,8	Slight economic downturn in 2019	Deflation	↓
12	AUSTRALIA	142	2,4	2,5	2,0	1,0	0,6	Sharp economic downturn in 2019	Deflation	↓
12	AUSTRALIA	143	2,4	2,0	1,0	0,0	-0,3	Risk of recession in 2019	Deflation	↓

South America

Country Nb	County name	Number	2016	2017	2018	2019	2020	End of cycle	Long term tendency	Indicator
16	MEXICO	111	3,3	3,1	4,0	5,0	5,5	Slight economic downturn in 2016	Optimistic	↑
16	MEXICO	112	2,8	2,3	3,0	4,0	5,0	Sharp economic downturn in 2016	Optimistic	↑
16	MEXICO	113	1,9	1,0	2,0	3,0	3,8	Risk of recession in 2016	Optimistic	↑
16	MEXICO	121	4,0	3,8	3,6	4,0	4,3	Slight economic downturn in 2017	Neutral	=
16	MEXICO	122	4,0	3,2	2,5	3,0	4,0	Sharp economic downturn in 2017	Neutral	=
16	MEXICO	123	4,0	3,2	0,6	2,0	3,0	Risk of recession in 2017	Neutral	=
16	MEXICO	131	3,5	3,5	3,3	3,1	3,8	Slight economic downturn in 2018	Stagnation	→
16	MEXICO	132	3,5	3,9	3,0	2,0	2,8	Sharp economic downturn in 2018	Stagnation	→
16	MEXICO	133	3,5	2,5	0,5	0,0	1,0	Risk of recession in 2018	Stagnation	→
16	MEXICO	141	2,5	3,5	2,5	2,3	2,3	Slight economic downturn in 2019	Deflation	↓
16	MEXICO	142	2,5	2,0	1,0	0,5	1,0	Sharp economic downturn in 2019	Deflation	↓
16	MEXICO	143	2,5	2,3	1,7	0,8	-0,5	Risk of recession in 2019	Deflation	↓
17	CHILE	111	3,0	2,8	3,0	3,3	3,7	Slight economic downturn in 2016	Optimistic	↑
17	CHILE	112	2,5	1,8	2,3	3,0	3,6	Sharp economic downturn in 2016	Optimistic	↑
17	CHILE	113	1,5	0,7	1,5	2,5	3,5	Risk of recession in 2016	Optimistic	↑
17	CHILE	121	3,4	3,2	3,4	3,6	3,4	Slight economic downturn in 2017	Neutral	=
17	CHILE	122	3,4	3,0	2,5	3,0	3,3	Sharp economic downturn in 2017	Neutral	=
17	CHILE	123	3,4	2,0	1,3	3,0	3,3	Risk of recession in 2017	Neutral	=
17	CHILE	131	3,0	3,2	3,0	2,8	3,4	Slight economic downturn in 2018	Stagnation	→
17	CHILE	132	3,0	2,9	2,0	2,0	3,0	Sharp economic downturn in 2018	Stagnation	→
17	CHILE	133	3,0	2,8	1,4	1,0	1,5	Risk of recession in 2018	Stagnation	→
17	CHILE	141	2,5	2,3	2,5	2,0	2,0	Slight economic downturn in 2019	Deflation	↓
17	CHILE	142	2,5	2,4	1,5	1,0	0,0	Sharp economic downturn in 2019	Deflation	↓
17	CHILE	143	2,5	2,0	1,0	0,0	0,3	Risk of recession in 2019	Deflation	↓
18	BRAZIL	111	3,7	3,6	4,0	4,6	5,2	Slight economic downturn in 2016	Optimistic	↑
18	BRAZIL	112	3,5	2,6	3,5	4,0	5,0	Sharp economic downturn in 2016	Optimistic	↑
18	BRAZIL	113	3,0	1,0	1,3	2,0	2,6	Risk of recession in 2016	Optimistic	↑
18	BRAZIL	121	3,7	3,5	3,7	3,9	4,2	Slight economic downturn in 2017	Neutral	=
18	BRAZIL	122	3,7	3,0	3,3	3,6	4,0	Sharp economic downturn in 2017	Neutral	=
18	BRAZIL	123	3,7	2,1	3,0	3,4	3,7	Risk of recession in 2017	Neutral	=
18	BRAZIL	131	3,3	3,3	3,0	3,6	3,6	Slight economic downturn in 2018	Stagnation	→
18	BRAZIL	132	3,3	3,2	3,0	2,4	3,0	Sharp economic downturn in 2018	Stagnation	→
18	BRAZIL	133	3,3	2,4	1,0	0,4	1,8	Risk of recession in 2018	Stagnation	→
18	BRAZIL	141	3,0	3,3	3,0	2,8	2,5	Slight economic downturn in 2019	Deflation	↓
18	BRAZIL	142	3,0	3,2	2,6	2,0	2,0	Sharp economic downturn in 2019	Deflation	↓
18	BRAZIL	143	3,0	2,2	1,4	0,0	-0,3	Risk of recession in 2019	Deflation	↓

A.2. Growth rates

World

Country Nb	County name	Number	2016	2017	2018	2019	2020	End of cycle	Long term tendency	Indicator
1	WORLD	111	2,8	3,0	3,3	3,6	3,9	Slight economic downturn in 2016	Optimistic	↑
1	WORLD	112	1,0	1,5	2,5	3,0	3,6	Sharp economic downturn in 2016	Optimistic	↑
1	WORLD	113	-1,0	1,0	2,0	2,4	3,5	Risk of recession in 2016	Optimistic	↑
1	WORLD	121	3,7	2,3	2,6	2,9	3,3	Slight economic downturn in 2017	Neutral	=
1	WORLD	122	3,7	1,0	2,0	2,5	3,5	Sharp economic downturn in 2017	Neutral	=
1	WORLD	123	3,7	-1,0	1,0	2,0	3,0	Risk of recession in 2017	Neutral	=
1	WORLD	131	3,0	3,2	2,3	2,6	2,9	Slight economic downturn in 2018	Stagnation	→
1	WORLD	132	3,0	3,3	1,0	2,0	3,0	Sharp economic downturn in 2018	Stagnation	→
1	WORLD	133	3,0	3,5	-1,0	1,0	2,0	Risk of recession in 2018	Stagnation	→
1	WORLD	141	2,5	2,6	2,8	2,4	3,0	Slight economic downturn in 2019	Deflation	↓
1	WORLD	142	2,5	3,0	3,3	1,0	2,0	Sharp economic downturn in 2019	Deflation	↓
1	WORLD	143	2,5	3,3	3,5	-1,0	1,0	Risk of recession in 2019	Deflation	↓

Europe

Country Nb	County name	Number	2016	2017	2018	2019	2020	End of cycle	Long term tendency	Indicator
2	GERMANY	211	1,6	1,5	1,8	1,8	2,0	Slight economic downturn in 2016	Optimistic	↑
2	GERMANY	212	1,3	1,0	1,5	1,7	1,9	Sharp economic downturn in 2016	Optimistic	↑
2	GERMANY	213	-0,6	0,5	1,0	2,1	2,0	Risk of recession in 2016	Optimistic	↑
2	GERMANY	221	1,3	1,1	1,3	1,4	1,6	Slight economic downturn in 2017	Neutral	=
2	GERMANY	222	1,3	0,5	0,8	1,0	1,3	Sharp economic downturn in 2017	Neutral	=
2	GERMANY	223	1,3	-1,0	0,7	1,0	1,3	Risk of recession in 2017	Neutral	=
2	GERMANY	231	0,8	1,4	1,0	1,1	1,3	Slight economic downturn in 2018	Stagnation	↗
2	GERMANY	232	0,8	1,0	0,3	1,0	1,3	Sharp economic downturn in 2018	Stagnation	↗
2	GERMANY	233	0,8	1,2	-1,0	0,8	1,2	Risk of recession in 2018	Stagnation	↗
2	GERMANY	241	0,5	0,8	1,4	1,2	1,2	Slight economic downturn in 2019	Deflation	↓
2	GERMANY	242	0,5	0,8	1,0	0,4	0,6	Sharp economic downturn in 2019	Deflation	↓
2	GERMANY	243	0,5	0,7	0,8	-1,0	0,0	Risk of recession in 2019	Deflation	↓
3	FRANCE	111	1,1	0,8	1,1	1,3	1,3	Slight economic downturn in 2016	Optimistic	↑
3	FRANCE	112	0,5	0,7	0,9	1,1	1,3	Sharp economic downturn in 2016	Optimistic	↑
3	FRANCE	113	-0,5	0,4	1,0	1,2	1,3	Risk of recession in 2016	Optimistic	↑
3	FRANCE	121	0,9	1,2	0,8	1,2	1,2	Slight economic downturn in 2017	Neutral	=
3	FRANCE	122	0,9	0,4	0,8	1,2	1,2	Sharp economic downturn in 2017	Neutral	=
3	FRANCE	123	0,9	-0,7	0,5	1,2	1,2	Risk of recession in 2017	Neutral	=
3	FRANCE	131	0,8	1,0	1,2	0,8	0,9	Slight economic downturn in 2018	Stagnation	↗
3	FRANCE	132	0,8	0,9	0,2	0,8	0,9	Sharp economic downturn in 2018	Stagnation	↗
3	FRANCE	133	0,8	0,9	-0,8	0,8	0,9	Risk of recession in 2018	Stagnation	↗
3	FRANCE	141	0,6	1,0	1,4	1,5	0,5	Slight economic downturn in 2019	Deflation	↓
3	FRANCE	142	0,6	0,7	0,8	0,0	0,5	Sharp economic downturn in 2019	Deflation	↓
3	FRANCE	143	0,6	0,6	0,6	-1,0	0,5	Risk of recession in 2019	Deflation	↓
4	ITALY	111	0,9	0,6	0,7	1,0	1,2	Slight economic downturn in 2016	Optimistic	↑
4	ITALY	112	0,5	0,6	0,7	0,9	1,2	Sharp economic downturn in 2016	Optimistic	↑
4	ITALY	113	-0,8	0,4	0,8	1,0	1,2	Risk of recession in 2016	Optimistic	↑
4	ITALY	121	1,2	0,7	0,6	0,8	1,0	Slight economic downturn in 2017	Neutral	=
4	ITALY	122	1,2	0,0	0,2	0,5	1,0	Sharp economic downturn in 2017	Neutral	=
4	ITALY	123	1,2	-1,0	0,0	0,5	1,0	Risk of recession in 2017	Neutral	=
4	ITALY	131	0,7	0,9	0,6	0,8	0,9	Slight economic downturn in 2018	Stagnation	↗
4	ITALY	132	0,7	0,7	0,3	0,6	0,9	Sharp economic downturn in 2018	Stagnation	↗
4	ITALY	133	0,7	0,8	-1,0	0,4	0,9	Risk of recession in 2018	Stagnation	↗
4	ITALY	141	0,6	0,8	0,8	0,6	0,7	Slight economic downturn in 2019	Deflation	↓
4	ITALY	142	0,6	0,7	0,7	0,0	0,5	Sharp economic downturn in 2019	Deflation	↓
4	ITALY	143	0,6	0,6	0,6	-1,1	0,3	Risk of recession in 2019	Deflation	↓
5	SPAIN	111	1,0	1,2	1,4	1,6	1,7	Slight economic downturn in 2016	Optimistic	↑
5	SPAIN	112	0,7	1,0	1,3	1,6	1,7	Sharp economic downturn in 2016	Optimistic	↑
5	SPAIN	113	-0,5	0,5	1,0	1,5	1,7	Risk of recession in 2016	Optimistic	↑
5	SPAIN	121	1,2	1,0	1,2	1,4	1,4	Slight economic downturn in 2017	Neutral	=
5	SPAIN	122	1,2	0,6	1,0	1,2	1,4	Sharp economic downturn in 2017	Neutral	=
5	SPAIN	123	1,2	-1,0	0,5	1,0	1,4	Risk of recession in 2017	Neutral	=
5	SPAIN	131	1,1	0,9	1,1	1,2	1,2	Slight economic downturn in 2018	Stagnation	↗
5	SPAIN	132	1,1	0,6	0,9	1,0	1,2	Sharp economic downturn in 2018	Stagnation	↗
5	SPAIN	133	1,1	1,1	-1,0	0,5	1,0	Risk of recession in 2018	Stagnation	↗
5	SPAIN	141	1,0	0,7	0,9	0,9	1,0	Slight economic downturn in 2019	Deflation	↓
5	SPAIN	142	1,0	0,3	0,6	0,8	1,0	Sharp economic downturn in 2019	Deflation	↓
5	SPAIN	143	1,0	0,9	0,9	-1,0	0,5	Risk of recession in 2019	Deflation	↓
6	GREECE	111	2,0	2,2	2,3	2,3	2,4	Slight economic downturn in 2016	Optimistic	↑
6	GREECE	112	0,6	0,8	1,0	1,2	1,4	Sharp economic downturn in 2016	Optimistic	↑
6	GREECE	113	-2,0	-1,0	0,0	0,5	1,2	Risk of recession in 2016	Optimistic	↑
6	GREECE	121	2,4	2,0	1,9	2,0	2,0	Slight economic downturn in 2017	Neutral	=
6	GREECE	122	2,4	1,0	1,2	1,4	1,4	Sharp economic downturn in 2017	Neutral	=
6	GREECE	123	2,4	-2,5	-0,5	0,5	1,0	Risk of recession in 2017	Neutral	=
6	GREECE	131	1,2	1,0	0,8	1,2	1,2	Slight economic downturn in 2018	Stagnation	↗
6	GREECE	132	1,2	0,8	0,3	0,7	1,0	Sharp economic downturn in 2018	Stagnation	↗
6	GREECE	133	1,2	1,0	-3,0	-1,5	0,0	Risk of recession in 2018	Stagnation	↗
6	GREECE	141	0,2	0,3	0,0	0,3	0,3	Slight economic downturn in 2019	Deflation	↓
6	GREECE	142	0,2	0,0	0,2	0,0	0,2	Sharp economic downturn in 2019	Deflation	↓
6	GREECE	143	0,2	0,0	-0,9	-3,0	-2,0	Risk of recession in 2019	Deflation	↓

English-speaking countries

Country Nb	County name	Number	2016	2017	2018	2019	2020	End of cycle	Long term tendency	Indicator
7	UK	111	2,0	2,2	2,0	1,8	2,0	Slight economic downturn in 2016	Optimistic	↑
7	UK	112	1,6	2,0	2,2	2,0	2,0	Sharp economic downturn in 2016	Optimistic	↑
7	UK	113	0,3	1,2	1,9	2,2	2,0	Risk of recession in 2016	Optimistic	↑
7	UK	121	2,2	1,9	2,2	2,2	2,0	Slight economic downturn in 2017	Neutral	=
7	UK	122	2,2	1,5	2,0	2,2	2,0	Sharp economic downturn in 2017	Neutral	=
7	UK	123	2,2	-0,2	1,5	1,8	2,0	Risk of recession in 2017	Neutral	=
7	UK	131	2,0	1,8	2,0	2,2	1,8	Slight economic downturn in 2018	Stagnation	→
7	UK	132	2,0	1,4	1,8	1,9	1,8	Sharp economic downturn in 2018	Stagnation	→
7	UK	133	2,0	2,4	-0,3	1,0	1,6	Risk of recession in 2018	Stagnation	→
7	UK	141	1,8	1,6	1,8	1,5	1,6	Slight economic downturn in 2019	Deflation	↓
7	UK	142	1,8	0,9	1,2	1,2	1,6	Sharp economic downturn in 2019	Deflation	↓
7	UK	143	1,8	2,0	1,8	-0,5	1,0	Risk of recession in 2019	Deflation	↓
8	USA	111	3,0	3,3	3,2	3,3	3,2	Slight economic downturn in 2016	Optimistic	↑
8	USA	112	1,5	1,9	2,5	3,1	3,0	Sharp economic downturn in 2016	Optimistic	↑
8	USA	113	-0,2	3,0	3,5	3,3	3,5	Risk of recession in 2016	Optimistic	↑
8	USA	121	3,3	3,0	3,3	3,5	3,0	Slight economic downturn in 2017	Neutral	=
8	USA	122	3,3	2,3	3,0	3,2	3,0	Sharp economic downturn in 2017	Neutral	=
8	USA	123	3,3	-0,5	1,5	2,8	3,0	Risk of recession in 2017	Neutral	=
8	USA	131	2,0	1,8	2,0	2,5	2,9	Slight economic downturn in 2018	Stagnation	→
8	USA	132	2,0	2,3	1,0	2,0	2,8	Sharp economic downturn in 2018	Stagnation	→
8	USA	133	2,0	2,5	-1,0	1,5	2,5	Risk of recession in 2018	Stagnation	→
8	USA	141	1,0	0,8	1,0	1,3	2,0	Slight economic downturn in 2019	Deflation	↓
8	USA	142	1,0	0,5	0,8	1,0	0,8	Sharp economic downturn in 2019	Deflation	↓
8	USA	143	1,0	1,2	1,5	-2,0	-0,5	Risk of recession in 2019	Deflation	↓

India

Country Nb	County name	Number	2016	2017	2018	2019	2020	End of cycle	Long term tendency	Indicator
13	INDIA	111	6,2	6,8	7,5	8,0	8,5	Slight economic downturn in 2016	Optimistic	↑
13	INDIA	112	5,1	6,0	7,0	8,0	8,5	Sharp economic downturn in 2016	Optimistic	↑
13	INDIA	113	3,0	5,0	6,0	7,6	8,3	Risk of recession in 2016	Optimistic	↑
13	INDIA	121	6,6	6,2	6,7	7,8	8,2	Slight economic downturn in 2017	Neutral	=
13	INDIA	122	6,6	4,0	5,5	7,5	8,0	Sharp economic downturn in 2017	Neutral	=
13	INDIA	123	6,6	1,0	3,0	5,0	7,0	Risk of recession in 2017	Neutral	=
13	INDIA	131	6,0	7,0	6,0	6,4	6,8	Slight economic downturn in 2018	Stagnation	→
13	INDIA	132	6,0	5,0	4,0	5,0	6,0	Sharp economic downturn in 2018	Stagnation	→
13	INDIA	133	6,0	7,0	-0,2	3,0	5,0	Risk of recession in 2018	Stagnation	→
13	INDIA	141	5,5	5,0	4,0	5,0	4,5	Slight economic downturn in 2019	Deflation	↓
13	INDIA	142	5,5	4,0	2,0	3,0	4,0	Sharp economic downturn in 2019	Deflation	↓
13	INDIA	143	5,5	5,0	3,0	-0,9	0,0	Risk of recession in 2019	Deflation	↓

Russia/Poland

Country Nb	County name	Number	2016	2017	2018	2019	2020	End of cycle	Long term tendency	Indicator
14	RUSSIA	111	0,8	1,2	1,0	0,9	1,0	Slight economic downturn in 2016	Optimistic	↑
14	RUSSIA	112	0,3	1,0	0,6	0,8	1,0	Sharp economic downturn in 2016	Optimistic	↑
14	RUSSIA	113	-0,6	0,6	1,0	1,3	1,5	Risk of recession in 2016	Optimistic	↑
14	RUSSIA	121	1,0	0,8	1,0	0,8	1,0	Slight economic downturn in 2017	Neutral	=
14	RUSSIA	122	1,0	0,4	0,6	0,7	0,9	Sharp economic downturn in 2017	Neutral	=
14	RUSSIA	123	1,0	-1,0	0,5	1,0	0,8	Risk of recession in 2017	Neutral	=
14	RUSSIA	131	0,8	0,5	0,6	0,8	0,6	Slight economic downturn in 2018	Stagnation	→
14	RUSSIA	132	0,8	0,2	0,4	0,5	0,5	Sharp economic downturn in 2018	Stagnation	→
14	RUSSIA	133	0,8	-0,2	-2,5	0,0	0,5	Risk of recession in 2018	Stagnation	→
14	RUSSIA	141	0,6	0,3	0,4	0,3	0,3	Slight economic downturn in 2019	Deflation	↓
14	RUSSIA	142	0,6	0,1	0,2	0,1	0,0	Sharp economic downturn in 2019	Deflation	↓
14	RUSSIA	143	0,6	0,0	-1,0	-3,5	-2,0	Risk of recession in 2019	Deflation	↓
15	POLAND	111	3,2	3,6	3,7	4,0	3,9	Slight economic downturn in 2016	Optimistic	↑
15	POLAND	112	2,1	2,8	3,5	3,9	4,0	Sharp economic downturn in 2016	Optimistic	↑
15	POLAND	113	-0,1	2,0	3,0	3,3	3,9	Risk of recession in 2016	Optimistic	↑
15	POLAND	121	3,5	3,2	3,5	3,8	3,8	Slight economic downturn in 2017	Neutral	=
15	POLAND	122	3,5	2,4	3,0	3,5	3,6	Sharp economic downturn in 2017	Neutral	=
15	POLAND	123	3,5	-0,5	1,5	3,0	3,5	Risk of recession in 2017	Neutral	=
15	POLAND	131	3,2	3,0	3,2	3,1	3,3	Slight economic downturn in 2018	Stagnation	→
15	POLAND	132	3,2	2,5	2,8	3,0	3,2	Sharp economic downturn in 2018	Stagnation	→
15	POLAND	133	3,2	2,0	-1,0	1,5	3,0	Risk of recession in 2018	Stagnation	→
15	POLAND	141	3,0	2,8	3,0	2,7	2,9	Slight economic downturn in 2019	Deflation	↓
15	POLAND	142	3,0	1,8	2,3	2,5	2,0	Sharp economic downturn in 2019	Deflation	↓
15	POLAND	143	2,0	1,5	1,0	-1,5	0,0	Risk of recession in 2019	Deflation	↓

Asia

Country Nb	County name	Number	2016	2017	2018	2019	2020	End of cycle	Long term tendency	Indicator
9	CHINA	111	6,3	6,8	6,5	6,0	5,8	Slight economic downturn in 2016	Optimistic	↑
9	CHINA	112	5,5	6,0	6,3	6,0	5,8	Sharp economic downturn in 2016	Optimistic	↑
9	CHINA	113	2,0	4,0	5,0	5,5	5,8	Risk of recession in 2016	Optimistic	↑
9	CHINA	121	7,0	6,5	6,8	6,0	5,5	Slight economic downturn in 2017	Neutral	=
9	CHINA	122	7,0	5,5	6,2	6,0	5,5	Sharp economic downturn in 2017	Neutral	=
9	CHINA	123	7,0	1,0	3,0	5,0	5,5	Risk of recession in 2017	Neutral	=
9	CHINA	131	6,7	6,5	6,0	5,0	5,0	Slight economic downturn in 2018	Stagnation	→
9	CHINA	132	6,5	5,5	5,8	5,4	5,0	Sharp economic downturn in 2018	Stagnation	→
9	CHINA	133	6,3	5,0	0,0	4,0	4,0	Risk of recession in 2018	Stagnation	→
9	CHINA	141	5,8	5,3	5,5	4,8	4,0	Slight economic downturn in 2019	Deflation	↓
9	CHINA	142	5,5	4,7	5,0	4,0	3,0	Sharp economic downturn in 2019	Deflation	↓
9	CHINA	143	4,0	3,5	3,0	0,0	0,2	Risk of recession in 2019	Deflation	↓
10	JAPAN	111	2,5	2,9	2,5	2,2	2,0	Slight economic downturn in 2016	Optimistic	↑
10	JAPAN	112	2,0	2,5	2,5	2,2	2,0	Sharp economic downturn in 2016	Optimistic	↑
10	JAPAN	113	-1,0	0,5	1,0	1,5	2,0	Risk of recession in 2016	Optimistic	↑
10	JAPAN	121	1,9	2,0	1,8	1,6	1,5	Slight economic downturn in 2017	Neutral	=
10	JAPAN	122	1,9	1,5	1,8	1,6	1,5	Sharp economic downturn in 2017	Neutral	=
10	JAPAN	123	1,9	-1,5	0,0	0,7	1,5	Risk of recession in 2017	Neutral	=
10	JAPAN	131	1,5	1,6	1,4	1,2	1,0	Slight economic downturn in 2018	Stagnation	→
10	JAPAN	132	1,5	1,5	0,8	1,0	1,0	Sharp economic downturn in 2018	Stagnation	→
10	JAPAN	133	1,5	1,5	-2,0	0,0	1,0	Risk of recession in 2018	Stagnation	→
10	JAPAN	141	1,3	1,1	1,2	1,0	0,8	Slight economic downturn in 2019	Deflation	↓
10	JAPAN	142	1,3	1,3	1,3	1,2	0,6	Sharp economic downturn in 2019	Deflation	↓
10	JAPAN	143	1,3	1,2	1,0	-2,5	0,2	Risk of recession in 2019	Deflation	↓
11	SOUTHKOREA	111	3,4	3,7	4,0	3,8	3,9	Slight economic downturn in 2016	Optimistic	↑
11	SOUTH KOREA	112	2,5	3,0	3,5	4,0	3,8	Sharp economic downturn in 2016	Optimistic	↑
11	SOUTH KOREA	113	1,0	2,0	3,0	4,0	4,0	Risk of recession in 2016	Optimistic	↑
11	SOUTH KOREA	121	3,6	3,2	3,4	3,6	3,8	Slight economic downturn in 2017	Neutral	=
11	SOUTH KOREA	122	3,6	2,7	3,2	3,5	3,7	Sharp economic downturn in 2017	Neutral	=
11	SOUTH KOREA	123	3,6	0,0	2,5	3,6	3,9	Risk of recession in 2017	Neutral	=
11	SOUTH KOREA	131	3,4	3,3	3,1	3,3	3,6	Slight economic downturn in 2018	Stagnation	→
11	SOUTH KOREA	132	3,4	3,3	2,4	2,9	3,4	Sharp economic downturn in 2018	Stagnation	→
11	SOUTH KOREA	133	3,4	3,8	-0,3	2,0	3,0	Risk of recession in 2018	Stagnation	→
11	SOUTH KOREA	141	2,5	2,4	2,3	2,0	2,5	Slight economic downturn in 2019	Deflation	↓
11	SOUTH KOREA	142	2,5	2,5	2,3	1,5	2,5	Sharp economic downturn in 2019	Deflation	↓
11	SOUTH KOREA	143	2,5	2,0	1,5	-0,5	2,0	Risk of recession in 2019	Deflation	↓
12	AUSTRALIA	111	2,8	3,0	3,2	3,4	3,6	Slight economic downturn in 2016	Optimistic	↑
12	AUSTRALIA	112	2,0	2,6	3,0	3,2	3,4	Sharp economic downturn in 2016	Optimistic	↑
12	AUSTRALIA	113	0,0	2,0	2,4	2,8	3,0	Risk of recession in 2016	Optimistic	↑
12	AUSTRALIA	121	3,1	2,9	3,1	3,0	3,2	Slight economic downturn in 2017	Neutral	=
12	AUSTRALIA	122	3,1	2,0	2,5	3,0	3,0	Sharp economic downturn in 2017	Neutral	=
12	AUSTRALIA	123	3,1	-1,0	1,5	2,4	2,8	Risk of recession in 2017	Neutral	=
12	AUSTRALIA	131	2,8	3,0	2,6	2,8	3,0	Slight economic downturn in 2018	Stagnation	→
12	AUSTRALIA	132	2,8	2,7	1,5	2,2	2,5	Sharp economic downturn in 2018	Stagnation	→
12	AUSTRALIA	133	2,8	0,5	-2,0	0,0	1,5	Risk of recession in 2018	Stagnation	→
12	AUSTRALIA	141	2,6	2,4	2,0	2,4	2,0	Slight economic downturn in 2019	Deflation	↓
12	AUSTRALIA	142	2,6	1,8	1,5	1,7	1,5	Sharp economic downturn in 2019	Deflation	↓
12	AUSTRALIA	143	2,0	1,0	0,0	-2,0	-0,5	Risk of recession in 2019	Deflation	↓

South America

Country Nb	County name	Number	2016	2017	2018	2019	2020	End of cycle	Long term tendency	Indicator
16	MEXICO	111	3,6	3,9	4,0	3,8	3,6	Slight economic downturn in 2016	Optimistic	↑
16	MEXICO	112	3,0	3,3	3,6	3,6	3,8	Sharp economic downturn in 2016	Optimistic	↑
16	MEXICO	113	1,0	2,0	3,0	4,0	3,7	Risk of recession in 2016	Optimistic	↑
16	MEXICO	121	3,9	3,6	3,8	3,9	4,0	Slight economic downturn in 2017	Neutral	=
16	MEXICO	122	3,9	2,9	3,3	3,7	3,8	Sharp economic downturn in 2017	Neutral	=
16	MEXICO	123	3,9	0,0	2,0	3,0	3,5	Risk of recession in 2017	Neutral	=
16	MEXICO	131	3,6	3,5	3,3	3,8	3,6	Slight economic downturn in 2018	Stagnation	→
16	MEXICO	132	3,6	3,2	2,5	3,0	3,4	Sharp economic downturn in 2018	Stagnation	→
16	MEXICO	133	3,6	2,7	-0,5	2,0	2,5	Risk of recession in 2018	Stagnation	→
16	MEXICO	141	3,3	3,3	3,0	3,2	3,0	Slight economic downturn in 2019	Deflation	↓
16	MEXICO	142	3,3	3,0	2,0	2,5	2,8	Sharp economic downturn in 2019	Deflation	↓
16	MEXICO	143	3,3	3,0	2,0	-1,0	1,8	Risk of recession in 2019	Deflation	↓
17	CHILE	111	3,0	3,3	3,6	3,9	4,2	Slight economic downturn in 2016	Optimistic	↑
17	CHILE	112	2,2	2,8	3,7	4,0	4,1	Sharp economic downturn in 2016	Optimistic	↑
17	CHILE	113	1,0	1,5	2,0	2,8	3,4	Risk of recession in 2016	Optimistic	↑
17	CHILE	121	3,2	3,0	3,4	3,8	3,7	Slight economic downturn in 2017	Neutral	=
17	CHILE	122	3,2	2,0	2,5	3,2	3,6	Sharp economic downturn in 2017	Neutral	=
17	CHILE	123	3,2	0,0	1,5	3,0	3,2	Risk of recession in 2017	Neutral	=
17	CHILE	131	3,0	2,8	2,5	3,2	3,1	Slight economic downturn in 2018	Stagnation	→
17	CHILE	132	3,0	2,5	2,0	2,8	3,0	Sharp economic downturn in 2018	Stagnation	→
17	CHILE	133	3,0	2,0	-0,5	1,0	2,8	Risk of recession in 2018	Stagnation	→
17	CHILE	141	2,7	2,5	2,7	2,4	2,6	Slight economic downturn in 2019	Deflation	↓
17	CHILE	142	2,7	2,6	2,6	1,8	2,5	Sharp economic downturn in 2019	Deflation	↓
17	CHILE	143	2,7	2,6	2,3	-1,0	1,0	Risk of recession in 2019	Deflation	↓
18	BRAZIL	111	1,8	2,0	2,1	2,5	2,8	Slight economic downturn in 2016	Optimistic	↑
18	BRAZIL	112	1,5	1,9	2,0	2,3	2,4	Sharp economic downturn in 2016	Optimistic	↑
18	BRAZIL	113	0,0	1,0	1,5	2,0	2,3	Risk of recession in 2016	Optimistic	↑
18	BRAZIL	121	2,4	2,2	2,4	2,5	2,6	Slight economic downturn in 2017	Neutral	=
18	BRAZIL	122	2,4	1,6	2,3	2,5	2,5	Sharp economic downturn in 2017	Neutral	=
18	BRAZIL	123	2,4	-0,3	1,5	2,0	2,4	Risk of recession in 2017	Neutral	=
18	BRAZIL	131	2,2	2,2	2,0	2,2	2,3	Slight economic downturn in 2018	Stagnation	→
18	BRAZIL	132	2,2	1,8	1,3	1,8	2,2	Sharp economic downturn in 2018	Stagnation	→
18	BRAZIL	133	2,2	2,2	-0,5	1,4	2,0	Risk of recession in 2018	Stagnation	→
18	BRAZIL	141	2,0	2,0	1,8	2,0	1,8	Slight economic downturn in 2019	Deflation	↓
18	BRAZIL	142	2,0	1,8	1,2	1,4	1,5	Sharp economic downturn in 2019	Deflation	↓
18	BRAZIL	143	2,0	1,8	1,5	-0,9	-1,0	Risk of recession in 2019	Deflation	↓

Bibliography

[ABE 14] ABE S., "Definition of abenomics", *Financial Times Lexicon*, 2014.

[ANG 02] ANG A., BEKAERT G., "International asset allocation with regime shifts", *Review of Financial Studies*, vol. 15, pp. 1137–1187, 2002.

[ANG 04] ANG A., BEKAERT G., "How do regimes affect asset allocation?", *Review of Financial Studies*, 2004.

[ART 99] ARTZNER P. *et al.*, "Coherent measures of risk", *Math. Finance*, vol. 9, 1999.

[BLA 72a] BLACK F., "Capital market equilibrium with restricted borrowing", *Journal of Business*, vol. 45, no. 3, pp. 444–455, July 1972.

[BLA 72b] BLACK F. *et al.*, *The Capital Asset Pricing Model: Some Empirical Tests*, Praeger, New York, 1972.

[BLA 14] BLANQUÉ P., *Essays in Postive Investent Management*, Economica, 2014.

[BLI 99] BLIN J., Arbitrage pricing theory, Applied Portfolio Technology, Working paper, APT, Inc., 1999.

[BOU 95a] BOUTHEVILLAIN K., MATHIS A., "Erreurs de prévisions: une prévision retrospective sur l'année 1993", *Economie et Statistique*, nos. 285–286, 1995.

[BOU 95b] BOUTHEVILLAIN K., MATHIS A., "Meilleur, en moyenne, que les prévisions individuelles, le consensus ne garantit pas contre des erreurs importantes", *Economie et Statistique*, no. 285–286, 1995.

[BOY 04] BOYER R., DEHOVE M., PILHON D., The Newsletter of the French Council of Economic Analysis, 2004.

[BRO 95] BROWN S.J., GOETZMANN W.N., "Performance persistence", *The Journal of Finance*, vol. 50, no. 2, pp. 679–698, June 1995.

[BUS 79] BUSINESS WEEK, "The Death of Equities?", *Business Week*, vol. 12.08.1979, p. 54, 1979.

[BUS 12] BUSINESS WEEK LONDON, *Financial Times*, vol. 24.05.2012, p. 13, 24 May 2012.

[CAM 14] CAMDESSUS M., "Toute crise prépare la suivante", *Mémoire*, 2014.

[CHA 00] CHANG E.C., CHENG J.W., KHORANA A., "An examination of herd behavior in equity markets: an international perspective", *Journal of Banking and Finance*, vol. 24, pp. 1651–1679, 2000.

[CHE 86] CHEN N.F., ROLL R., ROSS S.A. *et al.*, "Economic forces and the stock market", *The Journal of Business*, vol. 59, 1986.

[CHR 95] CHRISTIE W.G., HUANG R.D., "Following pied piper: do individual returns herd around the market?", *Financial Analysts Journal*, pp. 31–37, July 1995.

[CLE 09] CLEMENT-GRANDCOURT A., JANSSEN J., *Méthodes quantitatives en gestion des risques financiers et papillons noirs*, Hermes–Lavoisier, Paris, 2009.

[CLE 10] CLEMENT-GRANDCOURT A., JANSSEN J., *Gestion des risques financiers et papillons noirs: Méthodes qualitatives*, Hermes–Lavoisier, Paris, 2010.

[COA 08] COATES J.M., HERBERT J., "Endogenous steroids and financial risk taking on a London trading floor", *Proceedings of the National Academy of Sciences*, vol. 105, no. 16, pp. 6167–6172, 22 April 2008.

[DEH 03] DEHOVE M., *Crises financières: deux ou trois choses que nous savons d'elles – que nous apprend l'approche statistique des crises financières?*, CAE, 2003.

[DRA 12] DRAGHI M., Speech at the Global Investment Conference in London, European Central Bank, 2012.

[ELT 96] ELTON E.J., GRUBER M., BLAKE C., "The persistence of risk-adjusted mutual fund performance", *The Journal of Business*, vol. 69, no. 2, pp. 133–157, April 1996.

[EUR 14] THE EUROPEAN SECURITIES AND MARKETS AUTHORITY, MIFID II and MIFIR Directives, 2014.

[FAM 92] FAMA E.F., FRENCH K.R. *et al.*, "The cross-section of expected stock returns", *The Journal of Finance*, 1992.

[FIS 82] FISHBURN P.C., The Foundations of Expected Utility, Theory and Decision Library, 1982.

[FRA 11] FRANCK D., Dodd-Frank Wall Street Reform and Consumer Protection Act, H. R. 4173, Enacted by the Senate and House of Representatives of the United States of America in Congress assembled, 30 September 2011.

[FRA 12] FRAYSSE H., *Construction d'un générateur de scénarios économiques à sauts permettant la prise en compte de scénarios de crises*, Institut des actuaires, France, 2012.

[FRI 82] FRIEDMAN M., SCHWARTZ A.J., *Monetary Trends in the United States and United Kingdom: their Relation to Income, Prices, and Interest Rates, 1867–1975*, University of Chicago Press, Series NBER Books, 1982.

[FRY 07] FRYDMAN R., GOLDBERG M.D., *Imperfect Knowledge Economics: Exchange Rates and Risk*, Princeton University Press, 2007.

[FRY 12] FRYDMAN R., PHELPS E.S., *Rethinking Expectations: the Way Forward for Macroeconomics*, Princeton University Press, 2012.

[GEH 12] GEHIN W., *Modélisation des queues de distribution des rendements des actifs financiers et application à la mesure et à la gestion du risque de marché*, Institut des actuaires, France, 2012.

[GEO 92] GEOGHEGAN T.J., CLARKSON R.S., FELDMAN K.S. *et al.*, "Report on the Wilkie stochastic investment model", *Journal of the Institute of Actuaries*, vol. 119, pp. 173–228, 1992.

[GIA 03] GIANNITSAROU C., "Supply-side reforms and learning dynamics – heterogeneous learning", *Review of Economic Dynamics*, 2003.

[GOE 94] GOETZMANN W., IBBOTSON R., "Do winners repeat? Patterns in mutual fund behaviour", *The Journal of Portfolio Management*, vol. 9, no. 2, pp. 9–18, 1994.

[GRI 85] GRINBLATT M., TITMAN S., "Approximate factor structures: interpretation and implications for empirical tests", *The Journal of Finance*, vol. 40, pp. 1367–1373, 1985.

[GRI 89] GRINBLATT M., TITMAN S., *Portfolio Performance Evaluation: Old Issues and New Insights*, Oxford University Press, 1989.

[HAS 14] HASSARI S.A., RAJHI W., "Mimétisme et marchés financiers européens", *Bulletin français d'actuariat*, Paris, vol. 14, no. 27, pp. 29–49, January–June 2014.

[HEN 91] HENDRICKS D., PATEL J., ZECKHAUSER R., Non Rational Actors and Financial Market Behavior, National Bureau of Economic Research, 1991.

[HEN 93] HENDRICKS D., PATEL J., ZECKHAUSER R., "Hot hands in mutual funds: short-run persistence of relative performance, 1974-1988", *The Journal of Finance*, vol. 48, no.1, pp. 93–130, March 1993.

[HOE 98] HOEUNG S., "La réplication de cap en tant qu'outil de protection d'un portefeuille obligataire", *Journal of the Institute of Actuaries*, 1998.

[HUN 96] HUNTINGTON S., The Clash of Civilizations and the Remaking of World Order, Foreign Affairs, 1996.

[HUN 97] HUNTINGTON S., "The west and the rest of the world", *Prospect Magazine*, 20 February 1997.

[HWA 04] HWANG S., SALMON M., "Market stress and herding", *Journal of Empirical Finance*, pp. 585–616, April 2004.

[IPC 13] IPCC, Climate Change 2013: The Physical Science Basis – Summary for Policymakers, Observed Changes in the Climate System, IPCC AR5 WG1, 2013.

[JAN 09] JANG I., KIM D., *Macroeconomics of Catastrophic Event Risks: Research Comes of Age*, BAILLY R.O. (ed.), New York, 2009.

[JEN 68] JENSEN M.C., "The performance of mutual funds in the period 1945-1964", *The Journal of Finance*, vol. 23, no. 2, pp. 389–416, 1968.

[JUN 14] JUNCKER J.-C., Political Guidelines, European Parliament, 2014.

[KEM 10] KEMP M., *Extreme Events: Robust Portfolio Construction in the Presence of Fat Tails*, Wiley, 2010.

[KEY 36] KEYNES J.M., *The General Theory of Employment, Interest, and Money*, Palgrave Macmillan, 1936.

[KON 43] KONDRATIEFF, *Theory of Long Cycles*, MIT Press, 1943.

[LEH 87] LEHMANN B.N., MODEST D.M., "Mutual fund performance evaluation: a comparison of benchmarks and benchmark comparisons", *The Journal of Finance*, vol. 42, no. 2, pp. 233–265, June 1987.

[LIN 69] LINTNER J., "The aggregation of investors' diverse judgements and preferences in purely competitive markets", *Journal of Financial and Quantitative Analysis*, vol. 4, no. 4, pp. 347–450, 1969.

[LUC 75] LUCAS R.F., "An equilibrium model of the business cycle", *Journal of Political Economy*, vol. 83, pp. 1113–1144, 1975.

[MAC 14] MACRON E., Projet de loi pour la croissance et l'activité, Assemblée nationale, 2014.

[MAR 52] MARKOWITZ H., "Portfolio selection", *Journal of Finance*, vol. 7, no. 1, pp. 77–91, March 1952.

[MCC 95] MCCULLOCH J.R., *The Collected Works of J. R. McCulloch*, Routledge, 1995.

[MER 73] MERTON RC., "An intertemporal capital asset pricing model", *Econometrica*, vol. 41, no. 5, pp. 867–887, September 1973.

[MUT 61] MUTH J.F., "Rational expectations and the theory of price movements", *Econometrica*, vol. 29, no. 3, pp. 315–335, 1961.

[NAT 11] NATIONAL ACADEMY, *America's Climate Choices*, The National Academies Press, Washington, DC, 2011.

[NEU 47] NEUMANN V.J., MORGENSTERN, *The Theory of Games and Economic Behavior*, Princeton University Press, 1947.

[NIE 99] NIEMIRA M.P., LAHIRI K., MOORE G.D., "International application of Neftci's probability approach", LAHIRI K., MOORE G.D. (eds), *Leading Economic Indicators: New Approaches and Forecasting Records*, Cambridge University Press, 1999.

[RED 09] REDOULÈS O., L'économie mondiale en 2008: du ralentissement à la récession, L'économie française, 2009.

[ROG 09] ROGOFF K.S., REINHART C.M., *This Time is Different – Eight Centuries of Financial Folly*, Princeton University Press, 2009.

[ROS 76] ROSS S.A., "The arbitrage theory of capital asset pricing", *Journal of Economic Theory*, vol. 13, 1976.

[ROU 06] ROUBINI N., Jackson Hole, 7 September 2006.

[SHA 92] SHARPE W.F., "Asset allocation: management style and performance measurement: an asset class model can help make order out of chaos", *Journal of Portfolio Management*, pp. 7–19, 1992.

[SIN 14] SINN H.W., "The Euro Trap: on bursting bubbles, budgets and beliefs", *Farewell to the Euro?*, Oxford University Press, Munich, 31 July 2014.

[SOR 94] SORTINO F., PRICE L., "Performance measurement in a downside risk framework", *Journal of Investing*, vol. 3, no. 3, pp. 59–64, 1994.

[STO 01] STONE M.R., WEEKS M., Systemic Financial Crises, Balance Sheets, and Model Uncertainty, International Monetary Fund Working Paper, Washington, DC, vol. 162, 2001.

[TIM 99] TIMMERMANS S., *Sudden Death and the Myth of CPR*, Temple University Press, Philadelphia, 1999.

[WAL 99] WALTER C., "Aux origines de la mesure de performance des fonds d'investissement: Les travaux d'Alfred Cowles", *Histoire & Mesure*, vol. 14, pp. 163–197, 1999.

[WEI 14] WEIDMANN J., "The Euro Crisis is Not Yet Behind *Us*", *Spiegel online*, 2014.

[WIC 98] WICKSELL K., *Interest and Prices*, Ludwig von Mises Institute, 1898.

[WIC 01] WICKSELL K., *Lectures on Political Economy*, vol. 2, 1901.

[WOR 11] WORTHINGTON S., "Calibrating the real-world ESG to your view: giving credibility to judgement", *ERM Conference: Winning strategies in the Solvency II World*, pp. 1–41, 1 July 2011.

[WU 90] WU Y., YAO K.L., "Low frequency circular polarized soliton in a ferromagnetic system under a strong bias magnetic field", *Commun. Theor. Phys*, 1990.

ENSAE Working Papers

BAGHERY S., NOU V., SOREL K., Etude Statistique de la Gestion Benchmarquée des OPCVM, 2002.

CHAGNARD B., CHOUVIN, SUDRIÈS, Modélisation des performances des Hedge Funds de taux et des OPCVM de taux français, 2005.

GRAVELLINI B., HAMILOU, VOTTERO, Méthode de construction de portefeuilles de fonds de fonds et de fonds d'actions basée sur un CAPM à 4 moments, 2006.

LEPAGE F., POCHON M., Outils d'aide à la construction de portefeuilles de fonds de Hedge Funds et d'obligations basée sur un CAPM à 4 moments, 2007.

THAMEUR G., MIKOU N., CAPM à plusieurs moments sur deux périodes pour la gestion de portefeuilles de fonds d'actions et de fonds de hedge funds long-short, 2008.

CERRAJERO B., SCHMITT F., Construction de portefeuilles de fonds d'actions et de hedge funds long-short d'actions par une méthode CAPM améliorée, 2008.

FRILOUX D., FONGANG G., MEUNIER, Les indicateurs avancés, notamment de l'OCDE sont-ils en avance sur les marchés actions pour prévoir, notamment, les récessions et les crises?, 2012.

PAULET A., VU P., Utilisation pour la gestion de portefeuille de hedge funds long-short de générateurs de scénarios spécifiques, 2012.

CHABANNES A., HASSANI L., Générateurs de Scénarios Économiques et Applications, 2013.

LAZAR C., ZIZI S., Changement de générateur de scénarios à l'approche d'une crise, 2014.

MIKAELLIAN H., REVELLE, Générateur de Scenarii, 2012.

LAIRD B., VUILLEMIN, Backtests on Scenario Crisis Generator, 2014.

Abbreviations

ACF:	autocorrelation function
ALM:	asset and liability management
ARIMA:	autoregressive integrated moving-average
ARMA:	autoregressive moving-average
ATP:	arbitrage pricing theory
CAPEX:	capital expenditure
CAPM:	capital asset pricing model
CDS:	credit default swap
CIO:	Chief Information Officer
CLI:	composite leading indicator
CPPI:	constant proportion portfolio insurance
CVaR:	Conditional Value-at-Risk
ECB:	European Central Bank
ECRI:	Economic Cycle Research Institute
ENSAE:	*École nationale de la statistique et de l'administration économique*
ESG:	Economic scenario generator
ESMA:	European Securities and Markets Authority
ETF:	exchange-traded fund
FED:	Federal Reserve System

GARCH:	generalized autoregressive conditional heteroscedasticity
GDP ratio:	Gross Domestic Product
GNP ratio:	Gross National Product
IFRI:	French Institute of International Relations (*Institut Français des Relations Internationales*)
IKE:	imperfect knowledge economics
IMF:	International Monetary Fund
INSEE:	French National Institute of Statistics and Economic Studies (*Institut national de la statistique et des études économiques*)
LTROs:	long term refinancing operations
MAR:	Regulation (EU) No 596/2014 of the European Parliament and of the Council of 16 April 2014 on market abuse (market abuse regulation)
MiFID I:	Directive 2014/65/EU of the European Parliament and of the Council on markets in financial instruments and amending Directive 2002/92/EC and Directive 2011/61/EU
MiFID or MiFID I:	Markets in Financial Instruments Directive – Directive.2004/39/EC of the European Parliament and the Council
MIFID:	Markets in Financial Instruments Directive
MiFIR:	Regulation (EU) No 600/2014 of the European Parliament and of the Council on markets in financial instruments and amending Regulation (EU) No 648/2012
MIFIR:	Markets in Financial Instruments Regulation
MPT:	modern portfolio theory NATIXIS
OECD:	Organization for Economic Co-operation and Development
OFCE:	French Observatory of Economics (*Observatoire français des conjonctures économiques*)
OIS:	understanding overnight index swaps
ORSA:	own risk self-assessment
PACF:	partial autocorrelation function
PDC:	portfolio decarburization fund
PSOE:	*Partido Socialista Obrero Español*

SAREB:	company for the management of assets proceeding from restructuring of the banking system in Spain
SARIMA:	seasonal autoregressive integrated moving average process
SCR:	solvency capital requirement
TED:	T-Bill Eurodollar
TVaR:	Tail Value at Risk
UKIP:	The UK Independence Party
VaR:	Value at Risk

Index

Printed in the United States
By Bookmasters